STECK-VAUGHN

Building Strategies™ SCIENCE

Susan D. McClanahan

Judith Andrews Green

Series Reviewers

▼ **Dr. Pamela Taylor Hakim**
Curriculum Coordinator for
  Distance Learning Program
The University of Mississippi
Oxford, Mississippi

▼ **Bobby Jackson**
Director of Adult Education
Roane County Schools
Kingston, Tennessee

▼ **Betty J. Kimberling**
Director of Adult Basic Education
St. Joseph Adult Education Center
St. Joseph, Missouri

▼ **Faith McCaghy**
Area Literacy Director
Dakota County Literacy Projects
Lakeville, Minnesota

▼ **John Ritter**
Master Teacher
Oregon Women's Correctional Center
Salem, Oregon

STECK-VAUGHN
COMPANY
ELEMENTARY · SECONDARY · ADULT · LIBRARY

Acknowledgments

Staff Credits

Executive Editor:	Ellen Northcutt
Supervising Editor:	Carolyn M. Hall
Design Manager:	Donna M. Brawley
Electronic Production:	Shelly M. Knapp
Cover Design:	D. Childress
Electronic Cover Production:	Alan Klemp
Photo Editor:	Margie Foster
Editorial Development:	McClanahan & Company, Inc.

Photography

Cover: (geological formations) © Robin Smith/Tony Stone Images; (fossil) © Mark Gibson/The Stock Market; (insect) © Superstock; p.15 © Barry L. Runk/Grant Heilman; p.16 © Tony Freeman/PhotoEdit; p.17 © Runk/Schoenberger/Grant Heilman; pp.22, 31 © Custom Medical Stock Photo; p.41 © David Young-Wolff/PhotoEdit; p.42 Kansas Travel and Tourism; p.43 © Superstock; p.44 © Stan Schroeder/Animals Animals; p.46 The Bettmann Archive; p.47 © Zig Leszczynski/Animals Animals; p.48 © Miriam Austerman/Animals Animals; p.53 © Robert Ginn/PhotoEdit; p.57 Texas Highways; p.59 © Larry Ulrich/Tony Stone Images; p.64 © Richard Harrington/FPG; p.69 © Superstock; p.74 Reuters/Bettmann; p.76 © Phil Degginger/Earth Scenes; p.78 © Camera Hawaii, Inc.; p.80 © Tom Dietrich/Tony Stone Images; pp.82, 84 Reuters/Bettmann; p.86 © Aneal Vohra/Unicorn Stock Photos; p.87 Solid Waste Management; p.90 The Bettmann Archive; p. 91 © Charles E. Schmidt/Unicorn Stock Photos; p.93 © Tom McCarthy/PhotoEdit; p.94 Randal Alhadeff; p.98 © Runk/Schoenberger/Grant Heilman; p.104 © Peter Simon/Stock Boston; p.107 Randal Alhadeff; p.108 The Wine Institute; p.110 Randal Alhadeff; p.114 © Bob Kramer/Stock Boston; p.119 © Aneal Vohra/Unicorn Stock Photos; p.129 © Westlight/Japack; p.134 © Superstock; p.140 © Photri; p.142 © Superstock.

Illustration

David Griffin, Rusty Kaim, Joe Ruszkowski, Eric Zill
Maps by Maryland Cartographics, Inc.

ISBN 0-8114-6501-2

Building Strategies is a trademark of Steck-Vaughn Company.

Copyright © 1996 Steck-Vaughn Company

All rights reserved. No part of the material protected by this copyright may be reproduced in any form by any means, electronic or mechanical, including photocopying, recording, or by any information storage and retrieval system, without permission in writing from the copyright owner. Requests for permission to make copies of any part of the work should be mailed to: Copyright Permissions, Steck-Vaughn Company, P.O. Box 26015, Austin, Texas 78755.

Printed in the United States of America.

3 4 5 6 7 8 9 10 DBH 00 99 98 97

Contents

To the Learner 5
Check What You Know 7
Skills Preview Chart 14

Unit 1: Biology 15
Lesson 1: The Scientific Method 16
Lesson 1 Practice 19
Lesson 2: The Circulatory System 20
Lesson 2 Practice 23
Lesson 3: The Nervous System 24
Lesson 3 Practice 27
Strategies for Success: Reading a Diagram 28
Lesson 4: The Skin 30
Lesson 4 Practice 32
Lesson 5: Bones and Muscles 33
Lesson 5 Practice 35
Lesson 6: Bacteria and Viruses 36
Lesson 6 Practice 39
Strategies for Success: The Main Idea; The Implied
  Main Idea 40
Lesson 7: The Balance of Nature 42
Lesson 7 Practice 45
Lesson 8: Extinction 46
Lesson 8 Practice 49
Lesson 9: Plants 50
Lesson 9 Practice 52
Lesson 10: The Environment 53
Lesson 10 Practice 55
Strategies for Success: Classifying 56
Thinking and Writing 58

Unit 2: Earth Science 59
Lesson 11: The Atmosphere 60
Lesson 11 Practice 63
Lesson 12: Ocean Currents 64
Lesson 12 Practice 67
Lesson 13: Volcanoes 68
Lesson 13 Practice 71
Strategies for Success: Reading a Map 72
Lesson 14: Earthquakes 74
Lesson 14 Practice 77
Lesson 15: Hot Spots 78
Lesson 15 Practice 81

Lesson 16: Hurricanes. 82
Lesson 16 Practice 85
Lesson 17: Pollution and Recycling. 86
Lesson 17 Practice 89
Strategies for Success: Using Context Clues 90
Thinking and Writing 92

Unit 3: Chemistry. 93
Lesson 18: Matter All Around. 94
Lesson 18 Practice 97
Lesson 19: States of Matter 98
Lesson 19 Practice 101
Strategies for Success: Cause and Effect. 102
Lesson 20: The Conservation of Matter 104
Lesson 20 Practice 106
Lesson 21: Fermentation 107
Lesson 21 Practice 109
Lesson 22: Hard and Soft Water 110
Lesson 22 Practice 112
Lesson 23: Acid Rain. 113
Lesson 23 Practice 115
Strategies for Success: Reading Tables and Graphs 116
Thinking and Writing 118

Unit 4: Physics. 119
Lesson 24: Sound 120
Lesson 24 Practice 123
Lesson 25: Gravity 124
Lesson 25 Practice 127
Strategies for Success: Drawing Conclusions. 128
Lesson 26: Light 130
Lesson 26 Practice 132
Lesson 27: Lasers 133
Lesson 27 Practice 135
Strategies for Success: Predicting Outcomes. 136
Lesson 28: Nuclear Energy. 138
Lesson 28 Practice 141
Lesson 29: Physics Explains Why 142
Lesson 29 Practice 144
Thinking and Writing 145

Check What You've Learned 146
Skills Review Chart. 153

Glossary 154
Answers and Explanations 161

To the Learner

In Steck-Vaughn's *Building Strategies™ Science*, you will study biology, earth science, chemistry, and physics. You will also practice your reading skills as you use maps, charts, graphs, diagrams, and tables. This book includes the following features especially designed to help you develop your science skills.

Skills Inventories
- Before you begin work, take the *Check What You Know* skills inventory, check your own answers, and fill out the *Skills Preview Chart*. There you will see which skills you already know and which skills you need to practice in this book.
- After you finish the last practice page, take the *Check What You've Learned* skills inventory, check your answers, and fill out the *Skills Review Chart*. You'll see the great progress you've made. Save these inventories in a folder or portfolio.

Strategies for Success
In each unit *Strategies for Success* sections give you tips and practice for ways to improve your reading skills immediately as you read science content. Reviewing these strategies from time to time will help you work through the book.

Thinking and Writing
On the *Thinking and Writing* pages, you will apply your reading and science skills while writing your opinions on various interesting topics. Save these examples of your writing in a folder or portfolio and later review how much you have accomplished.

Practice Pages
A *Practice* page follows each lesson. These pages contain a variety of exercises such as vocabulary in context, classifying, drawing conclusions, and reading diagrams. Check your own work after finishing the exercises to find out how well you understood the lesson.

Extension
Information on topics related to the lesson appears periodically in *Extension* activities. These activities give interesting and helpful additional information. You'll have a chance to give your opinions here.

Glossary
The glossary at the back of the book lists and defines the science terms in *Building Strategies™ Science*. It also tells you the page number on which each term first appears.

Answers and Explanations
This section gives you the answers to all the questions so you can check your own work. Sample answers are usually given for open-ended questions that have no one right answer. Answers to multiple-choice questions also explain why one choice is correct and why the other possible answer choices are incorrect. Studying the explanations can help you sharpen your test-taking skills.

Tips for Your Success
There are a few tips to making learning easier. Try all these tips and decide which ones work best for you.

Hard Words. You don't have to know every word in this book to understand what you're reading. When you come to a hard word, keep on reading. The rest of the sentence or paragraph will probably help you figure out what the hard word means. Terms in bold letters in this book are explained in the margin. Also, use the *Glossary* to review the terms. If a hard word is not in the *Glossary*, look up the word in a dictionary.

Understanding New Subjects. When you are learning a new subject, such as gravity, understanding comes a little bit at a time. When you read something that seems very hard, put a question mark with your pencil by the part you don't understand. Keep reading to the end of the paragraph or lesson. Then reread the part you questioned; it will probably begin to make sense. Try to connect the information you are reading to the examples on the page.

Preview the Pages and Predict. Study the table of contents, the unit titles and lesson titles, the unit opening pages, and the pictures. Also read the definitions of important terms in the margins of the pages. All these study aids can help you predict what you are about to read and help you understand the information.

Check What You Know

Check What You Know will give you an idea of the kind of work you will be doing in this book. It will help you know how well you understand science content. It will also show you which reading skills you need to improve.

You will read passages, graphs, and maps followed by one or more multiple-choice questions. There is a total of 20 questions. There is no time limit.

Read each passage and question carefully. Fill in the circle for the best answer.

Questions 1-2 are based on the following paragraph.

There are three kinds of blood vessels: arteries, veins, and capillaries. The arteries carry blood away from the heart, and the veins carry blood back to the heart. The capillaries deliver the blood to the body's cells. The blood is always circulating in the body. For this reason, the heart and the blood vessels are known as the circulatory system.

1. Something that circulates probably

   Ⓐ contains several parts that work together.
   Ⓑ moves in a circle.
   Ⓒ has blood in it.
   Ⓓ looks like a blood vessel.

2. You can conclude that the purpose of the circulatory system is to

   Ⓐ clean the blood.
   Ⓑ move blood from one place to another.
   Ⓒ make blood.
   Ⓓ stop bleeding when you cut yourself.

Questions 3–5 are based on the following paragraph.

The planarian, a flatworm, has some interesting features. First, it can grow back missing parts. If a planarian's head or tail is cut off, it can grow a new head or tail. If a piece is cut out of the middle of the worm, that piece will grow both a head and a tail. Second, a planarian can reproduce asexually—without having any sexual contact. To do this, it pulls itself apart in the middle. Then each section grows back the missing part. The section without a head grows a head, and the section without a tail grows a tail.

3. To reproduce *asexually* means to reproduce

   Ⓐ by having sexual contact.
   Ⓑ without having sexual contact.
   Ⓒ by pulling apart.
   Ⓓ by being cut into pieces.

4. The main idea of the paragraph is that

   Ⓐ the planarian is a flatworm.
   Ⓑ when a planarian's head is cut off, it grows a new head.
   Ⓒ a planarian can pull itself apart.
   Ⓓ the planarian has some interesting features.

5. What would happen if a planarian were cut in half?

   Ⓐ The tail part would grow a head, and the head part would grow a tail.
   Ⓑ The tail part would grow a head, but the head part would die.
   Ⓒ The head part would grow a tail, but the tail part would die.
   Ⓓ Both halves of the planarian would die.

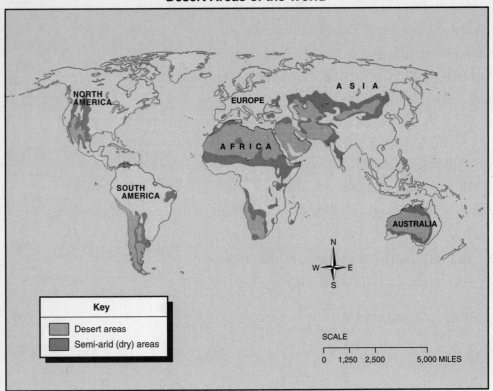

Desert Areas of the World

Questions 6–8 are based on the following map.

6. Which continent is almost all desert or semi-arid?

   Ⓐ North America
   Ⓑ South America
   Ⓒ Africa
   Ⓓ Australia

7. Which continent has no deserts?

   Ⓐ Asia
   Ⓑ Africa
   Ⓒ Europe
   Ⓓ South America

8. Where is the largest desert area located?

   Ⓐ Australia
   Ⓑ North America
   Ⓒ South America
   Ⓓ Africa

Questions 9–11 are based on the following passage.

If you stir instant coffee into a cup of hot water to make coffee, you are making a solution. You are dissolving something, the coffee, into something else, the water. Tea and most soft drinks are also solutions.

Solutions don't have to be liquid. Many metal objects are made of hard solutions called alloys. An alloy is made of two or more metals. These metals are melted and dissolved together. When the metal hardens, the result is a solution. Most of the coins we use are alloys. A nickel is only 25 percent nickel. The rest is copper. Even gold jewelry is an alloy—usually of gold and copper.

9. Which of the following is a solution?

    Ⓐ salt water
    Ⓑ pure gold
    Ⓒ mixed nuts
    Ⓓ gravel

10. Which of the following is an alloy?

    Ⓐ soda
    Ⓑ mineral water
    Ⓒ a penny
    Ⓓ air

11. What would be the result if you left a cup of sweetened instant coffee out on a table for several days?

    Ⓐ There would be nothing left in the cup.
    Ⓑ Only instant coffee powder would be left in the cup.
    Ⓒ Only water would be left in the cup.
    Ⓓ Both sugar crystals and instant coffee powder would be left in the cup.

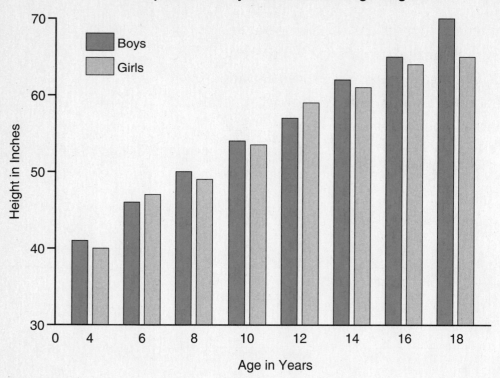

Questions 12–14 refer to the following graph.

12. At what age are girls taller than boys?

Ⓐ 10
Ⓑ 12
Ⓒ 14
Ⓓ 16

13. During what period do boys grow the most?

Ⓐ between 8 and 10
Ⓑ between 10 and 12
Ⓒ between 14 and 16
Ⓓ between 16 and 18

14. Suppose the graph also showed the height of boys and girls at age 2. What would probably be true of their height at age 2?

Ⓐ Boys and girls would be about the same height.
Ⓑ Boys would be much taller than girls.
Ⓒ Girls would be much taller than boys.
Ⓓ Height at age 2 cannot be measured exactly.

Questions 15–17 refer to the following paragraph and diagram.

A flat mirror reflects light at the same angle at which the light strikes it. The image appears to be the same size as the object. A concave mirror is curved inward like the inside of a bowl. Light is reflected toward a focal point and then spreads out. The image appears larger than the object. A convex mirror curves outward like the outside of a bowl. Light is reflected so that the image is smaller than the object.

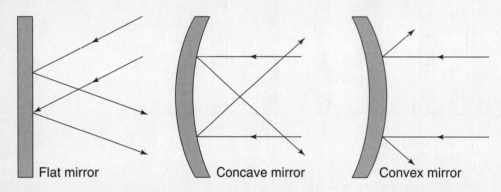

Flat mirror   Concave mirror   Convex mirror

15. Which mirror reflects light so that it does not spread out?

    Ⓐ flat mirror
    Ⓑ concave mirror
    Ⓒ convex mirror
    Ⓓ both flat and concave mirrors

16. Which mirror reflects light toward a central point?

    Ⓐ flat mirror
    Ⓑ concave mirror
    Ⓒ convex mirror
    Ⓓ both flat and concave mirrors

17. What is the main idea of the paragraph and diagram?

    Ⓐ There are three types of mirrors.
    Ⓑ The image in a flat mirror appears the same size as the object.
    Ⓒ No matter what type of mirror is used, an image appears smaller than the object.
    Ⓓ Concave mirrors are curved.

Questions 18–20 are based on the following paragraph.

Most air pollution comes from burning fuels like oil, gas, and coal. We burn fuels in factories and cars. We burn fuels to make electricity and to heat buildings. As a result, tiny particles of smoke, exhaust fumes, and other substances are in the air we breathe. They can damage people's health, cause acid rain, poison lakes and rivers, and destroy trees.

18. What is the main cause of air pollution?

    Ⓐ weather that stays the same for days
    Ⓑ burning fuels
    Ⓒ heating buildings
    Ⓓ poisoned lakes

19. What would probably happen if we reduced air pollution?

    Ⓐ There would be fewer cars on the road.
    Ⓑ Factories would burn less fuel.
    Ⓒ People would be healthier.
    Ⓓ The amount of electricity would increase.

20. What is the main idea of the paragraph?

    Ⓐ Air pollution does a lot of damage.
    Ⓑ Burning fuels causes air pollution.
    Ⓒ Air pollution can poison lakes.
    Ⓓ There are tiny particles in the air.

When you finish *Check What You Know*, check your answers on pages 161–162. Then complete the chart on page 14.

Check What You Know

The chart shows you which skills you need to study. Reread each question you missed. Then look at the appropriate pages of the book for help in figuring out the right answers.

Skills Preview Chart

Skills	Questions	Pages
The test, like this book, focuses on the skills below.	Check (√) the questions you missed.	Preview what you will learn in this book.
Reading a Diagram	___ 15 ___ 16	UNIT 1 ◆ Pages 15–58 Strategy for Success Pages 28–29
Finding the Main Idea	___ 4 ___ 17 ___ 20	UNIT 1 ◆ Pages 15–58 Strategy for Success Pages 40–41
Classifying	___ 9 ___ 10	UNIT 1 ◆ Pages 15–58 Strategy for Success Pages 56–57
Reading a Map	___ 6 ___ 7 ___ 8	UNIT 2 ◆ Pages 59–92 Strategy for Success Pages 72–73
Using Context Clues	___ 1 ___ 3	UNIT 2 ◆ Pages 59–92 Strategy for Success Pages 90–91
Cause and Effect	___ 11 ___ 18	UNIT 3 ◆ Pages 93–118 Strategy for Success Pages 102–103
Reading Tables and Graphs	___ 12 ___ 13	UNIT 3 ◆ Pages 93–118 Strategy for Success Pages 116–117
Drawing Conclusions	___ 2 ___ 19	UNIT 4 ◆ Pages 119–145 Strategy for Success Pages 128–129
Predicting Outcomes	___ 5 ___ 14	UNIT 4 ◆ Pages 119–145 Strategy for Success Pages 136–137

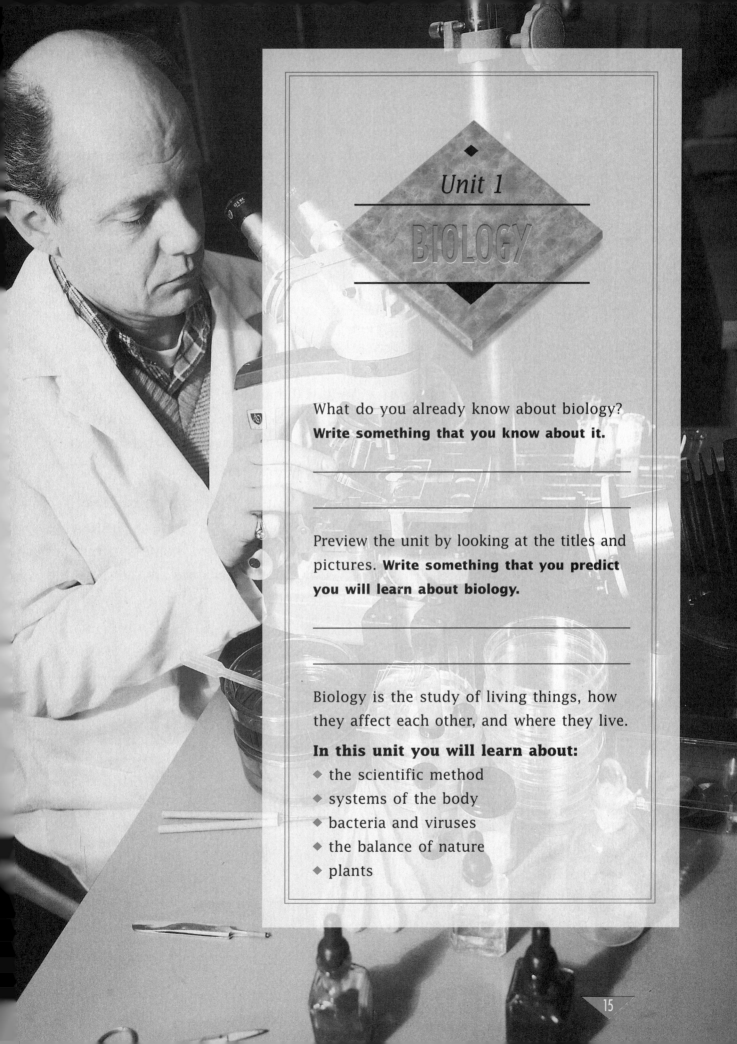

Unit 1
BIOLOGY

What do you already know about biology? **Write something that you know about it.**

_____

_____

Preview the unit by looking at the titles and pictures. **Write something that you predict you will learn about biology.**

_____

_____

Biology is the study of living things, how they affect each other, and where they live.

In this unit you will learn about:
- the scientific method
- systems of the body
- bacteria and viruses
- the balance of nature
- plants

Lesson 1

The Scientific Method

Have you ever eaten a home-grown tomato, picked right off the plant? If you have, you know that a home-grown tomato is much tastier than a supermarket tomato. A home-grown tomato is red, juicy, firm, and very tasty. In contrast, a supermarket tomato is often tough and mealy with a mild taste.

Supermarket tomatoes are bred to be firm so they will survive long truck hauls. They are also picked green, before they become sweet, so they will stay fresh for weeks.

Can a supermarket tomato be developed that tastes like a home-grown tomato? A scientist who asks such a question has a strategy for finding the answer. That strategy is called the **scientific method**. The scientific method is a way of getting information and testing ideas. It gives scientists a logical way to solve problems.

The first step of the scientific method is to **observe**. This means scientists look carefully at a problem. In the case of the tomato, scientists studied home-grown and supermarket tomatoes. They observed how these tomatoes are the same and how they are different. They observed what makes home-grown tomatoes taste so good. They observed what features of supermarket tomatoes make them easy to ship and store.

scientific method
A process for getting information and testing ideas.

observe
Watch and read to gather information about something.

Tomatoes fresh from the garden taste better than the ones that have been hauled a long distance to the supermarket.

Unit 1
16

The second step of the scientific method is to state the problem as a question. For the plant scientists, the question might be: How can we develop a tomato that can be picked when ripe yet keep for weeks?

The third step of the scientific method is to guess the answer to the question. This guess is not a wild guess. Instead, it is a good guess, based on knowledge and experience. It is called a **hypothesis**. Scientists working on the tomato problem came up with this hypothesis: There is a **gene** that causes the tomato to get soft when it is ripe. If the gene is inserted backward into new tomato plants, then the tomatoes will ripen without softening.

The fourth step of the scientific method is to **experiment**. Experiments must be designed to test whether the hypothesis is true. To test the backwards-gene hypothesis, scientists found and reversed the soft-when-ripe gene and inserted it into a group of tomato plants. They also grew other tomato plants without the reversed gene. Scientists performed these experiments on many varieties of tomatoes.

hypothesis
A good guess about the answer to a question.

gene
A microscopic part of a living thing that tells the living thing how to develop.

experiment
A method used to test a hypothesis.

Scientists studied home-grown tomatoes.

Lesson 1

conclusion
The decision on whether the hypothesis is true or false.

The fifth and last step of the scientific method is to draw a **conclusion**. From the results of experiments, scientists decide whether the hypothesis is true. Sometimes the hypothesis proves to be true; sometimes it is false. The conclusions that scientists reach often lead them to new observations and problems. The plant scientists who experimented with tomatoes concluded that their hypothesis was true. They could use genes to develop a tomato that can be picked close to ripeness and keep in the supermarket.

The following table summarizes the scientific method.

The Scientific Method

Step	Example
1. Observe.	1. Home-grown tomatoes, picked when ripe, taste better than supermarket tomatoes.
2. State the problem as a question.	2. How can we develop a tomato that can be picked when ripe yet keep for weeks?
3. Make a hypothesis.	3. There is a gene that causes the tomato to get soft when it is ripe. If the gene is inserted backward into new tomato plants, then the gene will cause them not to get soft when they are ripe.
4. Experiment.	4. Grow some tomatoes with the reversed gene and others without it.
5. Draw a conclusion.	5. Reversed genes can produce a tomato that can be picked when ripe and then stored.

Lesson 1

Practice

Vocabulary in Context ◆ Write the word that best completes each sentence.

1. The _____ method is a way that scientists solve problems.

2. A _____ is a guess, based on experience and knowledge, about the answer to a problem.

3. To test their ideas, scientists perform _____ .

> conclusion
> experiments
> hypothesis
> scientific

Finding Facts ◆ Choose the word or words that best complete each sentence. Fill in the circle for your answer.

4. Scientists study a problem and look at it carefully. This process is called
   - Ⓐ observation.
   - Ⓑ experimentation.
   - Ⓒ hypothesis.
   - Ⓓ conclusion.

5. After scientists experiment, they decide whether their hypothesis is true or false. They draw
   - Ⓐ an experiment.
   - Ⓑ a conclusion.
   - Ⓒ a scientific method.
   - Ⓓ an observation.

Sequence ◆ Write the five steps of the scientific method in the correct order.

6. _____

_____

_____

_____

_____

Check your answers on page 162.

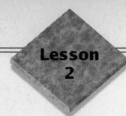

Lesson 2

The Circulatory System

Your heart is a pump. It is about as big as your fist and weighs less than a pound. It beats about 70 times a minute, over 100,000 times a day. It pumps blood all day, every day. The heart is always on duty. When you run, it speeds up. When you rest, it slows down.

Adults have between four and five quarts of blood in their bodies. The heart pumps the blood through the body so quickly that all of it passes through the heart once every minute. The blood is pumped to every part of the body through **blood vessels**. These blood vessels carry the blood throughout the body. There are about 100,000 miles of blood vessels in the body. If you put all your blood vessels in a line, they would stretch around the world four times.

Basically there are three kinds of blood vessels: **arteries**, **veins**, and **capillaries**. The arteries carry blood away from the heart and the veins carry blood back to the heart. The capillaries deliver the blood to the body's cells. The blood is always circulating around the body. For this reason, the heart and the blood vessels are known as the **circulatory system**.

In order for the body to live, all the cells need oxygen. The cells use the oxygen and all the nutrients in the blood. The arteries take fresh blood, rich in oxygen, all around the body for the body's cells to use. The arteries don't deliver the fresh blood directly to the cells. The arteries are like large pipes. They deliver the blood to smaller pipes, the capillaries. A capillary is so small that only one blood cell at a time can pass through its walls and feed the body's cells.

blood vessel
A tube that blood flows through.

artery
A blood vessel that carries blood away from the heart.

vein
A blood vessel that carries blood to the heart.

capillary
A very small blood vessel.

circulatory system
The heart and all the blood vessels.

The cells' wastes are fed back into capillaries. These capillaries pass the "used" blood on to the veins. The veins then take the blood to the liver, lungs, and kidneys. They clean wastes from the blood, which then returns to the heart. The heart pumps the blood to the lungs, where the heart receives a fresh supply of oxygen. From the lungs the "fresh" blood goes back to the heart. The heart then pumps the fresh blood around the body again.

The Circulatory System

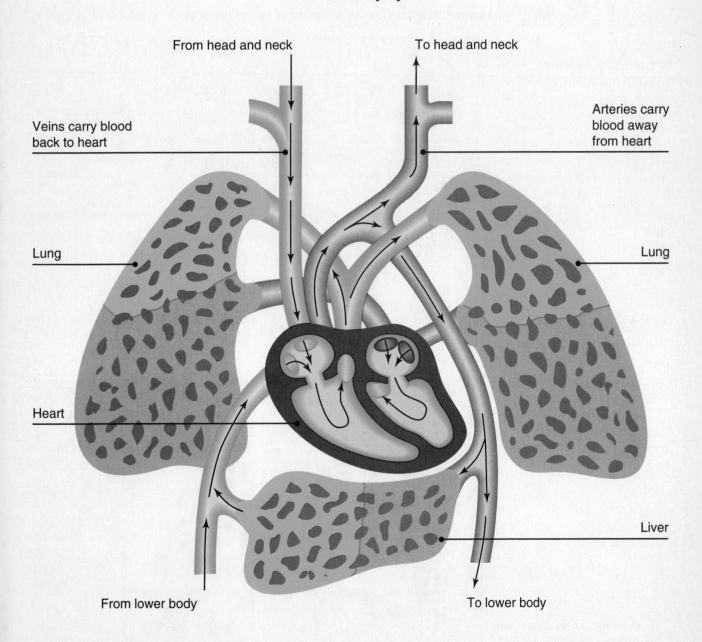

coronary artery disease
A disease that causes the arteries to become blocked.

cholesterol
A type of fat found in the human body and in animal foods.

heart attack
A condition in which part of the heart dies from lack of blood.

Diseases of the heart and circulatory system can be deadly. The most common heart disease is **coronary artery disease**. In this disease, the blood vessels leading to the heart become blocked with fat and **cholesterol**. As a result, the heart does not get enough blood and part of it may die. When this happens, a person has a **heart attack**. Each year, about 1,500,000 Americans have heart attacks. About one-third of them die.

Although heart attacks cannot always be prevented, there are things you can do to reduce your risk of getting coronary artery disease.

Reducing the risk of heart disease:
- Eat foods that are low in fat and cholesterol. This means eating fewer eggs, as well as less cheese, meat, milk, and ice cream. Eat more bread, rice, pasta, fruits, and vegetables.
- Exercise regularly. Walking, running, swimming, skating, and other aerobic exercises are helpful. They work by lowering the amount of fat and cholesterol in the blood. Aerobic exercise also strengthens the heart.
- If you are a smoker, give up smoking. Heavy smokers have twice the risk of heart disease as nonsmokers.

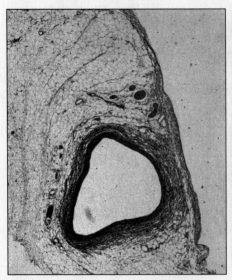
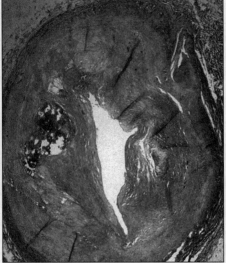

A healthy artery and an artery clogged with fat

Lesson 2

Practice

Vocabulary in Context ◆ **Write the word that best completes each sentence.**

1. A blood vessel that carries blood to the heart is called a _____ .

2. A blood vessel that carries blood away from the heart is called an _____ .

3. A tiny blood vessel that delivers fresh blood to the cells is called a _____ .

artery
capillary
cholesterol
vein

Finding Facts ◆ **Choose the words that best answer each question. Fill in the circle for your answer.**

4. Where does blood go after it passes through the arteries?
   - Ⓐ to the capillaries
   - Ⓑ to the veins
   - Ⓒ to the heart
   - Ⓓ to the liver, lungs, and kidneys

5. Where does fresh blood go from the lungs?
   - Ⓐ to the capillaries
   - Ⓑ to the veins
   - Ⓒ to the heart
   - Ⓓ to the liver, lungs, and kidneys

Finding Facts ◆ **Write your answer below.**

6. List three things people can do to lower their risk of getting coronary artery disease.

_____

_____

_____

Check your answers on page 162.

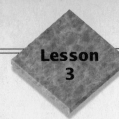

Lesson 3

The Nervous System

brain
The part of the body that controls all its activities.

cerebrum
The part of the brain in which thinking occurs.

cortex
The part of the cerebrum that thinks and stores information.

cerebellum
The part of the brain that coordinates muscle activities.

brain stem
The part of the brain that controls automatic life processes.

What part of you controls automatic activities like blinking and breathing? What body part makes sense of what you see, hear, touch, taste, and smell? What body part gives instructions to your muscles when you walk down the street? What part of you thinks thoughts and feels feelings? If you answered the **brain**, you are right.

Your brain is the most complex part of your body. It weighs only three pounds, yet it controls everything you do. The brain is divided into three major parts.

1. The largest part, the **cerebrum**, is responsible for thinking and deciding to act. Humans have the largest and most complex cerebrums of any animals. This is why humans are more intelligent than other animals. The outer layer of the cerebrum, the **cortex**, is where most information is stored.

2. The **cerebellum** is one-eighth the size of the cerebrum. It makes the muscles work together.

3. The **brain stem** regulates basic life functions like breathing and heartbeat. It also controls how alert you are.

The Human Brain

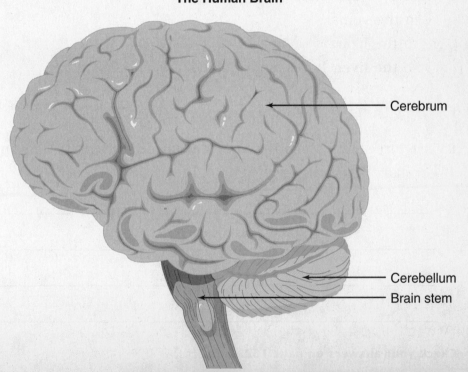

Unit 1
24

nervous system
All the nerves in the body, including the brain.

spinal cord
A part of the body that allows messages to travel between the brain and the body.

Your brain is the control center for sending and processing information. The **nervous system** consists of the brain and the nerves that go through the **spinal cord** and from there to every part of the body.

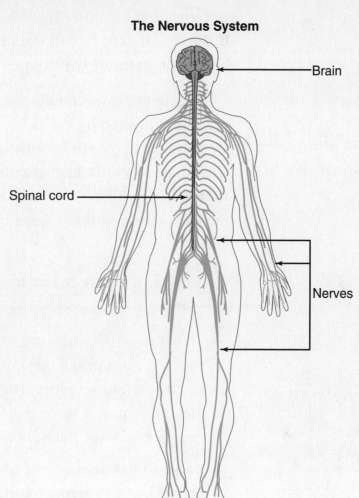

The Nervous System

Nerves throughout your body carry information to and from the spinal cord and brain. For example, the nerves in your fingertips send information about what you touch to the spinal cord and brain. Nerves that go to organs such as the kidneys and heart are also part of the nervous system.

The brain and the spinal cord receive, process, and respond to information. For example, you might see a dog, recognize her, and pat her head. This information is processed in the brain. On the other hand, you might touch a hot pot handle and drop the pot. This is a **reflex**. Nerves in your spinal cord make you drop the pot. Nerves in your brain make you feel the pain in your hand.

reflex
A quick response caused by nerves in the spinal cord.

Lesson 3

The nervous system is fragile. If the brain, nerves, or spinal cord is hurt, the results can be serious. When part of the brain is damaged, a person loses one or more abilities. For example, a person who had an accident may not be able to remember anything that happened before the accident. If the spinal cord is cut, the person cannot feel and move below the point of the injury.

Fortunately, the nervous system is well protected. The brain is protected by the hard skull. The spinal cord is inside a strong bony covering. Both the brain and spinal cord are surrounded by a fluid that softens shocks. Even the major nerves are well covered. Most of them are below layers of muscle or behind bony areas like the eye socket.

Extension
Wearing a Helmet

Because brain damage is so serious, many cities and states require people to protect themselves when taking part in certain activities. For example, motorcycle riders may be required to wear helmets. People may also be required to wear helmets when riding bikes.

Do you think that people should be required to wear helmets when playing sports such as football or when riding a bicycle? Write your opinion below.

_____

_____

_____

_____

_____

Check your answers on page 163.

Lesson 3

Practice

Vocabulary in Context ♦ **Write the word that best completes each sentence.**

1. The control center of the nervous system is the _____ .

2. The brain _____ is the part of the brain that regulates basic activities like breathing and heartbeat.

3. Thinking takes place in the _____ .

4. The _____ makes sure the muscles are working together.

brain
cerebellum
cerebrum
reflex
stem

Finding Facts ♦ **Choose the words that best answer each question. Fill in the circle for your answer.**

5. For which of the following activities is the brain responsible?
   Ⓐ providing energy for the cells in the body
   Ⓑ pumping blood throughout the body
   Ⓒ making sense of what you hear, see, touch, taste, and smell
   Ⓓ removing waste from the blood

6. What is a reflex?
   Ⓐ a complicated thought
   Ⓑ the result of a brain injury
   Ⓒ part of the spinal cord
   Ⓓ a quick response

7. Which of the following protects the nervous system?
   Ⓐ the heart
   Ⓑ the bones
   Ⓒ the nerves
   Ⓓ the eyes

Check your answers on page 163.

Strategies for SUCCESS

Reading a Diagram

Diagrams can help you understand science. A diagram is a drawing of the parts of something or how it works. The drawing usually has words to explain what you are looking at.

When you look at a diagram, first figure out the main idea. The main idea is the general subject of the diagram. The main idea is given in the title or caption. The main idea of this diagram is the nervous system.

The Nervous System

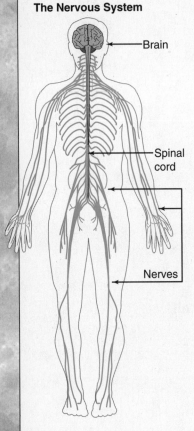

— Brain

— Spinal cord

— Nerves

> ❖ **STRATEGY:** **Look at the diagram. Ask yourself: What is the main idea?**
>
> 1. Look at the title or caption. It tells you the main idea of the diagram.
>
> 2. Look at the whole drawing. Ask yourself: What is this drawing about?

Exercise 1: **Look at the diagram below. Circle the title. Then write the main idea.**

_____

_____

_____

_____

Muscles of the Upper Arm

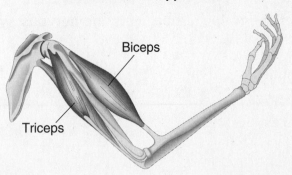

Biceps

Triceps

Unit 1
28

Once you understand the main idea, look carefully at the details. Figure out what the parts of the diagram show. Often there are word labels that tell what the parts are or what they do.

In the diagram of the nervous system, the labels are *brain*, *spinal cord*, *nerves*. Each label shows you an important detail.

> ❖**STRATEGY:** **Look at the diagram. Ask yourself: What are the important details of this diagram?**
>
> **1.** Look at the parts of the diagram. Ask yourself: What do these parts show?
>
> **2.** Read the labels. They tell you the important details.

Exercise 2: Look at the diagram of the brain shown below, which shows what different parts of the cerebrum do. Underline the labels. Then write two details that the diagram shows.

_____

_____

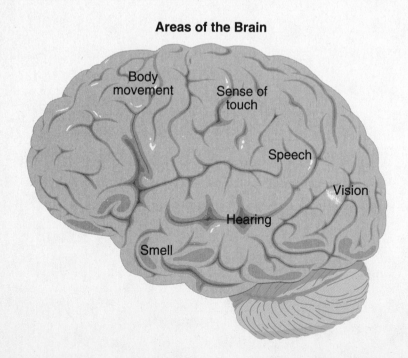

Areas of the Brain

Check your answers on page 163.

Lesson 3
29

Lesson 4

The Skin

organ
A part of the body that does a particular job. Organs include the heart and the skin.

cell
The basic part of any living thing. All plants and animals are composed of one or more cells.

germ
A microscopic organism, especially bacteria, that can cause disease.

infection
The presence in the body of a disease.

If someone asked you to list the important **organs** of the human body, you would probably mention the brain, the heart, and maybe the lungs or kidneys. You probably wouldn't mention the skin. But the skin is very important. Skin is the largest organ of the body. We have about 18 square feet of skin. It weighs about six pounds. Without it, we would die.

The skin has many jobs. Because it is almost waterproof, the skin keeps the body from drying out. This is important because the body is 65 percent water. Also, the skin helps keep the body at the right temperature. When we are hot, we sweat through our skin. This cools us down. The skin also has nerve **cells** that recognize pain, heat, and pressure and tell the body what is happening around it. The skin produces vitamin D when it is exposed to the sun. Vitamin D is necessary for strong, healthy bones. Perhaps most important of all, the skin protects the body from dirt and **germs**. The skin protects us from deadly **infections**.

The Layers of the Skin

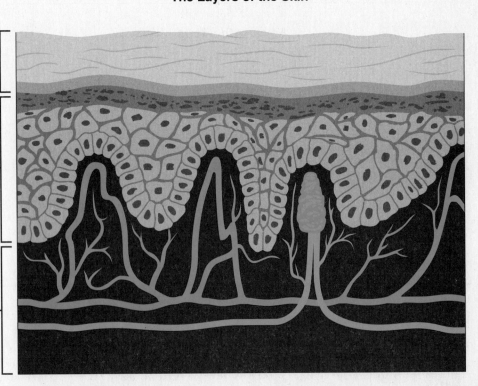

Outer layer of dead skin

Layer where new skin cells are made

Blood vessels and nerves

Unit 1

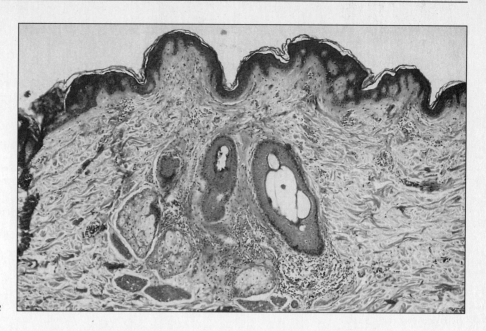

Skin cells as seen through a microscope

We usually take the skin for granted. But when we get badly burned, we suddenly realize how important skin is. Consider the case of Anthony, a young man from Brooklyn, New York. Anthony was working in a factory when a few sparks fell into a chemical drum. There was an explosion. In a few seconds, Anthony lost 40 percent of the skin from his neck, chest, stomach, back, and face.

Thirty years ago, a person with burns like Anthony's had only a 50 percent chance of surviving. Burn victims died then because without skin the body dried out rapidly. It couldn't recover quickly enough for the victim's survival. In addition, without skin to protect the body, infections set in. They spread throughout the body and killed the burn victim.

graft
Take skin from one part of the body and put it over an injured area.

Today we have drugs to fight infection, and we know how to replace body fluids. We also know how to **graft** skin from one part of the body to another. Grafting is often necessary when the skin is so badly damaged that it would not grow back normally. For these reasons, we now can save people even in cases where 75 percent of the skin has been burned.

As you can see, the skin is just as important as the heart, brain, or other organs. Keeping skin healthy helps you to stay healthy.

Lesson 4

Practice

Vocabulary in Context ♦ **Write the word that best completes each sentence.**

1. The skin, heart, lungs, and brain are all examples of _____ .

2. The skin provides protection against _____ , which cause disease.

3. Skin has nerve _____ that feel pain, heat, and pressure.

| cells |
| germs |
| infections |
| organs |

Finding Facts ♦ **Choose the words that best answer each question. Fill in the circle for your answer.**

4. Which of the following does the skin do?

   Ⓐ keeps the body upright
   Ⓑ keeps the body from drying out
   Ⓒ keeps the body strong
   Ⓓ protects the body from a poor diet

5. Why are bad burns over a large area of the body so serious?

   Ⓐ It is impossible to graft skin over a large area.
   Ⓑ Burn victims can die from drying out or infection.
   Ⓒ The skin has nerve cells.
   Ⓓ Burned skin can grow back, but it takes a long time.

Finding Facts ♦ **Write your answer below.**

6. List three things doctors do for victims of bad burns.

   _____

   _____

   _____

   _____

Check your answers on page 163.

Lesson 5

Bones and Muscles

Your bones support and protect your body from the inside. The 206 bones of the human skeleton carry the body's weight. With the help of muscles, your skeleton allows you to move. It also protects your organs. The skull covers the brain, and the ribs protect the heart and lungs.

Bones are very light. In fact, the skeleton of a 160-pound person weighs only about 29 pounds. Bones are light for two reasons. First, they have many tiny holes that are pathways for blood vessels and nerves. Second, the long bones of the arms and legs are made like hollow tubes. The **marrow** inside these bones is light.

Despite their light weight, bones are extremely strong. Your skeleton can hold up more than 5 times its weight. Pound for pound, for example, your thighbones are stronger than reinforced concrete. This strength comes from the materials that make bone. About half of a bone's weight is made of hard minerals like calcium and phosphorus. A quarter of its weight is made up of a protein fiber, and the rest is mostly water. The minerals and fiber are cemented together to make bones strong.

Your bones are strong, but to hold you up they need help. **Ligaments** hold the bones together. Muscles move bones or keep them in place. Muscles are attached to bones by **tendons**. Bones, ligaments, muscles, and tendons work together to help you stand and move.

Muscles move bones by pulling them. Because muscles can only pull, not push, they work in pairs. One muscle of the pair contracts (shortens), pulling the bone toward it. At the same time, the other muscle of the pair relaxes, allowing the bone to move. To make the bone move the other way, the first muscle relaxes and the second one contracts.

marrow
The material inside bones that produces blood cells.

The Skeleton

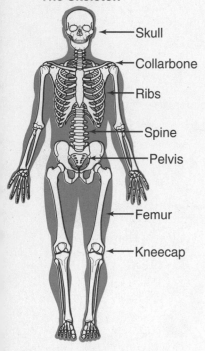

- Skull
- Collarbone
- Ribs
- Spine
- Pelvis
- Femur
- Kneecap

ligament
A tough, elastic band that holds two bones together.

tendon
A tough, elastic band that attaches a muscle to a bone.

The Muscles

- Trapezius
- Deltoid
- Triceps
- Gluteus maximus
- Hamstrings

joint
A place where bones are linked together.

For example, when you bend your arm at the elbow, you contract the biceps muscle on the front of the upper arm. The triceps muscle on the back of the upper arm relaxes. When you straighten your arm, the triceps contracts and the biceps relaxes.

Your muscles and bones move in thousands of ways. The body is flexible because of the way bones are linked together. The human body contains several types of **joints** that permit different kinds and ranges of movement.

Joints:

- Hinge joints, like the elbow and knee, move back and forth. They cannot move from side to side.
- Plane joints, like those in the spine, permit small gliding movements.
- Ball-and-socket joints, like the shoulder and hip, move freely in almost all directions.
- Saddle joints, like those of the ankle and thumb, allow movement in two directions at right angles.

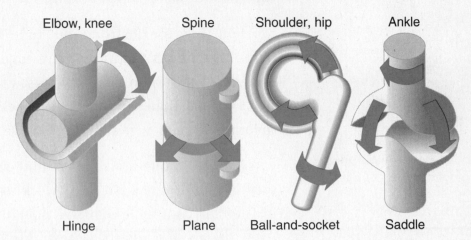

Four Types of Joints

Elbow, knee — Hinge
Spine — Plane
Shoulder, hip — Ball-and-socket
Ankle — Saddle

Lesson 5

Practice

Vocabulary in Context ◆ **Write the word or words that best complete each sentence.**

1. Blood cells are produced by material called _____ in the center of certain bones.

2. A _____ links two bones together at a place called a _____ .

3. A _____ connects a muscle to a bone.

joint
ligament
marrow
skeleton
tendon

Finding Facts ◆ **Choose the words that best answer each question. Fill in the circle for your answer.**

4. Why are bones so light?

   Ⓐ Bones are made like hollow tubes.
   Ⓑ Bones are made almost entirely of calcium.
   Ⓒ Bones are made of a protein fiber.
   Ⓓ Bones contain water.

5. Why are bones so strong?

   Ⓐ The marrow in bones makes them strong.
   Ⓑ Minerals and protein fiber are cemented together.
   Ⓒ Tendons attaching them to muscles make them strong.
   Ⓓ Ligaments that connect bones make them strong.

Reading a Diagram ◆ **Write your answer below.**

6. A fracture is a break in a bone. This diagram shows fractures in the bones of the lower arm. What is the main idea of the diagram?

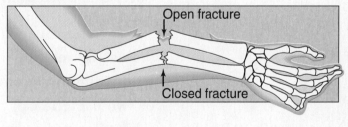

_____

_____

_____

Check your answers on pages 163–164.

Lesson 6

Bacteria and Viruses

bacteria
Tiny one-celled organisms. They live in water, soil, air, plants, animals, and people.

virus
A tiny organism, smaller than bacteria, that can reproduce only by using living cells.

All around you, there are tiny organisms called **bacteria** and **viruses**. They are in the air we breathe, the water we drink, and the food we eat. Although most bacteria and viruses do not harm people, many can cause disease. Disease-causing bacteria and viruses are often called germs.

Bacteria are divided by shape into three main groups which are listed below.

Shapes of Bacteria
- bacilli—rod-shaped bacteria
- spirilla—coiled bacteria
- cocci—sphere-shaped bacteria

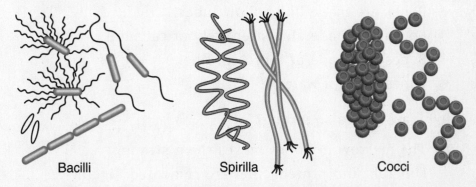

Bacilli   Spirilla   Cocci

Different kinds of bacteria cause different diseases. For example, bacilli cause salmonella food poisoning, tuberculosis, and whooping cough. Spirilla cause Lyme disease and syphilis. Cocci cause strep infections and acne.

Bacteria cause disease in two ways. Some bacteria cause disease by multiplying quickly in the body. They reproduce by dividing in two over and over again. In a few hours, one or two bacteria can reproduce into millions. This is how tuberculosis is caused.

toxin
A poisonous chemical.

The second way bacteria cause disease is by giving off chemicals called **toxins**. The toxins damage parts of the body, sometimes causing death. Botulism is a deadly type of food poisoning caused by a toxin.

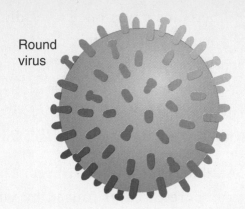 

Round virus

Bullet-shaped virus

Viruses come in different shapes.

Viruses are much smaller than bacteria. However, like bacteria, viruses come in different shapes. Viruses that cause the common cold, for example, are round. The virus that causes rabies is shaped like a bullet. Other viruses are shaped like rods or cubes. All viruses have an outer layer, called a protein coat, which protects the material inside.

Viruses cause disease by taking over living cells. A virus enters a living cell and uses it to make more copies of the virus. This damages or destroys the cell. As each new virus enters a new cell, the disease spreads through the body. Most viruses attack a particular type of cell. For example, the flu virus attacks the cells in the lining of the lungs, throat, and nose.

The body has natural defenses against disease-causing bacteria and viruses. When a germ enters the body, the **immune system** responds. The immune system also produces **antibodies** especially for that type of germ. The antibodies destroy the germs. **White blood cells** are sent to help destroy the germs.

Scientists keep looking for new ways to treat diseases caused by bacteria and viruses. When a disease is caused by bacteria, the body's immune system can be helped by **antibiotics**. These drugs destroy bacteria. However, bacteria can change over time. For example, every few years, a new form appears of *Streptococcus* bacteria, the bacteria that causes a sore throat. When new forms of bacteria develop, some antibiotics may no longer be effective.

immune system
The body's system for fighting disease-causing germs.

antibody
A chemical made by the immune system to destroy a particular germ.

white blood cell
A cell that can surround and kill germs.

antibiotic
A drug that stops the growth and reproduction of bacteria.

Lesson 6

Tuberculosis was once a deadly disease. Once scientists developed antibiotics, this disease could be cured. It was not very common in the 1960s. However, new forms of the tuberculosis bacteria have developed. These forms are not affected by many antibiotics. That's one reason tuberculosis is again spreading in the United States.

Antibiotics can't kill viruses. There are no other drugs that can cure diseases caused by viruses. But we do have drugs that relieve symptoms while your body fights the virus. For example, a cold tablet helps stop your sneezing and runny nose. It does not attack the cold virus.

Immunization can protect against many types of disease-causing bacteria and viruses. **Vaccines**, which use a weakened or dead form of the germ, are given by injection or by mouth. They cause the immune system to make antibodies. If the immunized person later gets that germ, he or she already has antibodies to fight it. Children are usually immunized against several diseases, which are listed below. As a result, most children in the United States no longer get these illnesses.

immunization
The process of creating resistance to particular germs.

vaccine
A substance that makes the body produce antibodies against a particular germ.

> **Childhood immunizations:**
> - polio
> - diphtheria, pertussis (whooping cough), tetanus (DPT)
> - measles, mumps, rubella (MMR)
> - chicken pox

Unfortunately, scientists have not yet been able to develop vaccines for many other diseases. Colds, for example, are caused by over 100 different types of viruses. If we had a vaccine for one or two of those viruses, it would not protect against the other types of viruses. Also, scientists have not developed a vaccine for HIV, the virus that leads to AIDS. The only defense against these viruses is to prevent becoming infected.

Lesson 6

Practice

Vocabulary in Context ◆ **Write the word or words that best complete each sentence.**

1. Disease-causing _____ and _____ are often called germs.

2. Some bacteria cause disease by giving off _____, or poisons.

3. _____ are chemicals produced by the immune system to fight particular germs.

antibiotics
antibodies
bacteria
toxins
viruses

Finding Facts ◆ **Choose the words that best answer the question. Fill in the circle for your answer.**

4. What are antibiotics used for?

   Ⓐ to give immunity against a disease
   Ⓑ to ease the symptoms of a disease
   Ⓒ to stop the growth and spread of bacteria
   Ⓓ to prevent the reproduction of viruses

Reading a Diagram ◆ **The diagram below shows how a virus causes a cell to make new viruses. Write your answers below.**

5. What happens to the cell at the end of this process?

   _____

6. What detail in the diagram helped you answer this question?

   _____

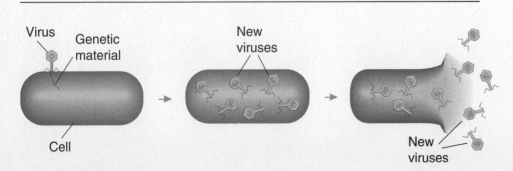

Check your answers on page 164.

The Main Idea

All the sentences in a paragraph usually relate in some way. In other words, there is one main idea. The main idea is often stated in one of the sentences of the paragraph. It is often in the first or last sentence. Knowing the main idea will help you understand the whole paragraph better.

> ❖**STRATEGY:** **Find the sentence that tells the main idea.**
>
> **1.** Read the whole paragraph.
>
> **2.** Find one sentence that sums up all the other sentences.

Exercise 1: Read the paragraph below. Underline the third sentence that gives the main idea.

    If someone asked you to list the important organs of the human body, you would probably mention the brain, the heart, and maybe the lungs or kidneys. You probably wouldn't mention the skin. But the skin is as important as the other organs. You have about 18 square feet of skin. Skin is the largest organ of the body. Without it, you would die.

Exercise 2: Read the paragraph below. Underline the sentence that gives the main idea.

    There are three kinds of blood vessels: arteries, veins, and capillaries. The arteries carry the blood away from the heart. The veins carry blood back to the heart. The capillaries deliver the blood to the body's cells.

The Implied Main Idea

Sometimes the main idea isn't stated in one of the sentences of a paragraph; it is implied. The sentences of the paragraph are still related. You can use the sentences to figure out what the main idea is.

Read the paragraph below. What is the main idea?

> The body's enemies—bacteria and viruses—are in the air we breathe. They're in the water we drink. They're in the food we eat. They're on the outside as well as on the inside of our bodies.

The paragraph mentions several different places where you can find bacteria and viruses. The main idea of the paragraph is: Bacteria and viruses are found almost everywhere.

> ❖**STRATEGY: Reread the paragraph for details. Then think about how the details are related.**
>
> 1. Ask yourself: How do all the sentences relate to each other? This is your clue.
>
> 2. Decide what main idea all the sentences add up to. This is the unstated main idea.

Exercise 3: Read the following paragraph. Write the main idea below.

> One of the skin's jobs is to keep the body from drying out. Skin helps keep the body at the right temperature. For example, when you're warm, sweating cools you off. Skin also protects the body from dirt and germs. When you get a cut, you lose some of that protection.

_____

_____

Check your answers on page 164.

Lesson 6

Lesson 7

The Balance of Nature

in balance
Stable.

ecosystem
A community where plants, animals, and climate are related and in balance.

Everything in nature exists **in balance**, a very complex and delicate balance. Plants are part of an **ecosystem**. Only a certain number of trees can grow in a forest. There is only so much light, so much space, and so much water. If there are too many trees, the strongest will crowd out the weakest until a balance is reached.

Animals are part of an ecosystem, too. Only a certain number of deer can live in a forest because there is only so much food for the deer to eat. If there is a warm winter, more deer than usual will stay alive. If more deer live through the winter, more young will be born that spring. More deer in the forest means less food for each one. Many of the deer will die because there is not enough food to support all of them. The following spring there will be fewer deer than there were before. This cycle happens naturally.

Even balanced ecosystems can change over time. For example, two hundred years ago, the Great Plains covered more than 220,000 square miles. It went from Canada to Texas and from Nebraska to the Great Lakes. The Great Plains was a tallgrass prairie ecosystem. Plant life included tough grasses and ferns and very few trees. Many kinds of animals lived in the ecosystem as well. The best known of these was the buffalo.

About 60 million buffalo used to roam the Great Plains.

Fire played an important role in Great Plains. It kept the Great Plains in balance. Fires burned many patches of prairie. Prairie plants have deep roots that can survive fire, so new plants grew right away. These young green plants attracted buffalo, who grazed on them. After eating all the plants in an area, the buffalo moved on. They left behind stalks that dried out, becoming good fuel for the next fire.

Grasses are the main plants of the plains.

When European settlers arrived, they damaged the ecosystem of the Great Plains in two ways. First, farmers plowed up the tough prairie grass so they could plant crops. Second, hunters killed buffalo for meat and fur. By 1900, fewer than 1,000 buffalo were left of the 60 million that had roamed the Plains. Fewer than 20,000 square miles of tallgrass prairie were left. The balance of the Great Plains ecosystem had been destroyed.

Today **conservation** groups are trying to restore small areas of tallgrass prairie. For example, in Oklahoma, a herd of 300 buffalo now grazes in part of a 36,000-acre tallgrass prairie **preserve**. In other parts of the Great Plains, small areas of tallgrass prairie are being restored. The buffalo **population** has risen to about 150,000.

conservation
Keeping something safe.

preserve
An area set aside for wildlife.

population
The number of a group that live in one place.

species
A group of plants or animals that are alike and can produce young together.

An ecosystem can become unbalanced in other ways. What if a new **species** of animal is introduced into an area? The ecosystem will change. This happened when American turtles were shipped from Florida to southern France to be sold as pets. However, many escaped or were let go when they got too big. Some of these turtles grow to a length of two feet. In the wild, the American turtles compete for food and space with the French turtles. Since the American turtles are much bigger, they are driving out the French turtles. As a result, the ecosystem in southern France is changing. It will reach a new balance in time.

Turtles from Florida are disturbing the ecosystem of southern France.

Extension
Ecosystems Around You

Ecosystems do not have to be wilderness areas. A city, suburb, or farm is an ecosystem. Ecosystems do not have to be large, either. Ponds, puddles, backyards, playgrounds, parks, or streets are all ecosystems.

Pick an ecosystem where you live. Describe the plants and animals there.

_____

_____

_____

_____

_____

_____

_____

_____

Unit 1

Check your answers on page 164.

Lesson 7

Practice

Vocabulary in Context ◆ Write the word that best completes each sentence.

1. An _____ is a community of related plants, animals, and climate.

2. All the buffalo on the Great Plains make up a group called _____.

3. A land area called a _____ is set aside to save part of an ecosystem.

ecosystem
population
preserve
species

Finding the Main Idea ◆ Choose the words that best answer each question. Fill in the circle for your answer.

4. What is the main idea of the second paragraph on page 42?

   Ⓐ Only a certain number of deer can live in a forest.
   Ⓑ Animals are part of balanced ecosystems.
   Ⓒ If there is a warm winter, more deer stay alive.
   Ⓓ If there are too many deer, the youngest survive.

5. What is the main idea of the first paragraph on page 44?

   Ⓐ American turtles are sold as pets in France.
   Ⓑ Many pet turtles escape or are let go into the wild.
   Ⓒ An ecosystem can become unbalanced when a new species is introduced.
   Ⓓ The American turtles are driving the smaller French turtles from the ecosystem.

Finding Facts ◆ Choose the words that best answer the question. Fill in the circle for your answer.

6. Why do prairie grasses grow soon after a fire?

   Ⓐ Their long roots survive fires.
   Ⓑ Their seeds grow best after a fire.
   Ⓒ The young green plants don't burn.
   Ⓓ The stalks left behind by buffalo don't burn.

Check your answers on page 164.

Lesson 8

Extinction

hairy mammoth
A large animal that is no longer alive. Mammoths looked like large, fur-covered elephants.

herbivore
An animal that eats only plants.

extinct
No longer in existence.

fossil
The remains of a once-living thing.

Until 10,000 years ago, **hairy mammoths** roamed northern Asia, Europe, and North America. Their long fur protected them from the cold. Larger than elephants, hairy mammoths grew to 12 feet tall with tusks up to 16 feet long. They weighed as much as three or four cars. Despite their large size, mammoths did not hunt animals. They were **herbivores**.

Mammoths had several natural enemies. Saber-toothed tigers and wolves hunted mammoths. However, their most dangerous enemies were humans. Prehistoric humans hunted mammoths for their meat, fur, and bones.

Another problem for mammoths was the weather. Over time, the earth's climate got colder. Mammoths' fur could not protect them from the colder weather. Eventually, hairy mammoths became **extinct**.

People today find many mammoth **fossils**. Most fossils are bones or tusks. More than 50,000 mammoth tusks have been found. Even whole animals, frozen solid, have been discovered in the Arctic.

Hairy mammoths once roamed northern Europe, Asia, and North America.

Mammoths are not the only creatures that have died out. Many species have become extinct. Dinosaurs ruled the earth for millions of years but are all extinct today. Many kinds of plants that lived at the same time as dinosaurs are extinct, too. The saber-toothed tigers, who hunted the mammoths, are also extinct.

subspecies
A group of plants or animals that look different but can produce young together.

Tigers of many kinds have existed for at least a million years. Tigers lived in Europe, the Americas, and Asia. Just 100 years ago there were eight **subspecies**, or breeds, of tigers. Now there are only five. The other subspecies are extinct.

Tigers are now found in the wild only in Asia. In 1900, in India alone there were probably 40,000 Bengal tigers. Now there are fewer than 2,500. There are only a few hundred Sumatran tigers and a few dozen Javan tigers. In all, there may be only 4,000 or 5,000 wild tigers left.

Asia is the only part of the world where tigers live in the wild.

Why have tigers been dying out? There are just two reasons. Until recent years, big-game hunters traveled to India or Africa to hunt wild animals. Big-game hunting has decreased due to new laws. However, in the early 1900s, thousands of Bengal tigers were hunted and killed.

habitat
The place where an animal or plant lives.

carnivore
An animal that eats only meat.

The second reason there are fewer tigers is human population growth. In 1891 there were about 280 million people in India. One hundred years later there were about 834 million. As the human population increased, people moved into jungles and changed them to farmlands. Many small animals died because their **habitat** was destroyed. These small animals were the main food for tigers, who are **carnivores**. Lack of living space also contributed to the decrease in the tiger population.

Lesson 8

endangered species
Any type of animal in danger of extinction.

Jaguars, wild cats that are smaller than tigers, are also carnivores. They eat wild and farm animals, fish, frogs, and alligators. Jaguars live in forests or open deserts; they can climb trees and swim. Jaguars can be found from Arizona in the United States to Argentina in South America. Like the tiger, the jaguar is an **endangered species**. Because of hunting and loss of their habitat, jaguars are now rare in most areas. In Argentina, for example, fewer than 200 wild jaguars are left.

As wilderness habitats have become smaller, zoos have become more important in saving endangered species. Workers in some zoos breed animals like tigers to keep them from becoming extinct. Some zoos breed animals so they can be trained to live in the wild. For example, the San Diego Zoo is training an endangered species of monkey called the lion-tailed macaque. Once the monkeys have been trained, they will be released in a wild area of India, their original home.

The lion-tailed macaque is being trained to live in the wild.

Another endangered species being bred at the San Diego Zoo is the California condor. Condors are the largest birds on earth. From tip to tip, their wings are 9 to 10 feet long. These birds were almost extinct when the zoo captured the few that were left to protect them and help take care of their young. For several years, the zoo has been breeding the condors. Recently, several were let go in the mountains of California.

Helping one species, like the condor, sometimes helps many other species. For example, 57 species of rare or endangered plants and animals live in the same habitat as the condor. By helping to save the condor's habitat, people help preserve the environment. Then other plants and animals that share the habitat have a better chance to survive.

In 1992, 1,200 species of animals were listed as endangered. Of these, 170 live in the United States. Laws now protect many of these animals and their habitats.

Lesson 8

Practice

Vocabulary in Context ◆ **Write the word that best completes each sentence.**

1. _____ species no longer exist at all.

2. _____ species may soon die out if we don't protect them.

3. _____ are animals that eat only meat.

> carnivores
> endangered
> extinct
> subspecies

Finding the Main Idea ◆ **Choose the words that best answer each question. Fill in the circle for your answer.**

4. What is the main idea of the first paragraph on page 47?

   Ⓐ Tigers are found all over the world.
   Ⓑ Several tiger subspecies became extinct in the last 100 years.
   Ⓒ Tigers were once common, but now they are rare.
   Ⓓ There are five subspecies of tigers today.

5. What is the main idea of the first paragraph on page 48?

   Ⓐ The jaguar is an endangered species.
   Ⓑ Jaguars are carnivores.
   Ⓒ Jaguars are found from Arizona to Argentina.
   Ⓓ There are fewer than 200 jaguars in Argentina.

Finding Facts ◆ **Choose the words that best answer each question. Fill in the circle for your answer.**

6. Why did mammoths become extinct?

   Ⓐ The animals they ate died out.
   Ⓑ The plants they ate died out.
   Ⓒ Saber-toothed tigers became extinct.
   Ⓓ The earth's climate became colder.

7. What is being done to save endangered animals?

   Ⓐ Zoos are breeding them.
   Ⓑ Laws protect their habitats.
   Ⓒ Laws protect them from hunting.
   Ⓓ all of the above

Check your answers on page 164.

Lesson 9

Plants

All animals, including humans, depend on plants to get energy needed to live. Green plants make their own food by using the energy of the sun. Animals can't get energy this way. Instead animals eat those plants and use their energy. Then plant-eating animals are eaten by other animals. So even when you eat a steak, you are getting energy from plants.

All animals need oxygen to live. Plants use our waste gas, carbon dioxide, and then give off oxygen into the air. We also get wood for houses, furniture, and fuel from plants. Plants are also the source of many medicines. Cotton, linen, and rayon clothing come from plants.

When you walk in a park or in the woods, you are surrounded by plants. There are over 300,000 species of plants in the world. Some are so small that you can't see them. Others, like giant sequoia trees, are so big that you can drive a car through a tunnel cut in the trunk. To better understand the great variety of plants, scientists divide them into groups.

- **Algae** can be tiny with only one cell. Some algae, such as seaweeds, are much larger. Most algae live in water.
- **Mosses and liverworts** are small plants that have no roots, stems, or leaves. They cover the ground like a rug.
- **Ferns** grow in shady places. they are larger than mosses and have roots, stems, and leaves. They do not have flowers or seeds.
- **Seed plants** are plants that reproduce by forming seeds. They also have roots, stems, and leaves. The seeds of evergreens are protected inside a core. In flowering plants, like apple trees, the seeds are protected inside the **fruit**.

algae
The simplest green plants without roots, stems, or leaves.

mosses and liverworts
Small green plants that grow in damp places.

fern
Green plants with roots, stems, and leaves but no seeds.

seed plants
Green plants with roots, stems, leaves, and seeds.

fruit
The part of a plant that contains seeds.

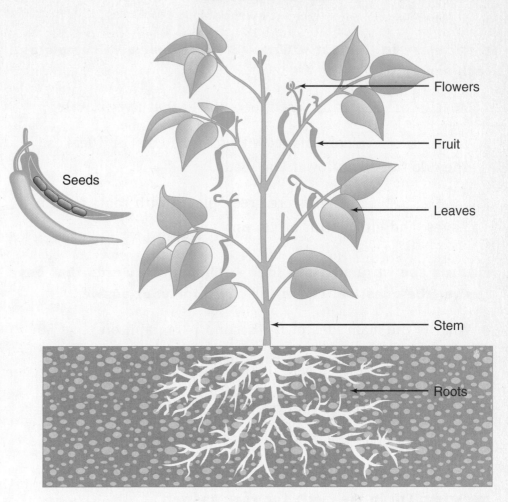

Parts of a Seed Plant

The parts of a seed plant work together to keep the plant alive and to make more plants. Roots hold the plant in the ground. They draw water and minerals from the soil up into the stem. The stem supports the leaves, flowers, and fruit. The stem also carries water and minerals to the upper parts of the plant. The leaves are the plant's factory. They absorb energy from the sun, produce sugar for food, and give off oxygen into the air. The flowers of the plant produce seeds, which are protected by a fruit or a cone.

Seed plants are the most common type of plants. More than half of the plant species are seed plants. Most plants grown in gardens or as crops for food are seed plants. Carrots and beets are the roots of seed plants. Corn, peas, and lima beans are seeds. Peaches, squash, and tomatoes are fruits. Foods don't have to be sweet to be fruits. Any part of a plant that has a seed inside is a fruit.

Lesson 9

Practice

Vocabulary in Context ◆ **Write the word that best completes each sentence.**

1. Most _____ are green plants that live in water.

2. _____ are plants with roots, stems, and leaves. They do not have flowers or seeds.

3. _____ plants are green plants with roots, stems, leaves, and either fruit or concs.

| algae |
| ferns |
| mosses |
| seed |

Finding the Implied Main Idea ◆ **Choose the words that best answer the question. Fill in the circle for your answer.**

4. What is the main idea of the second paragraph on page 50?

    Ⓐ Plants clean the air we breathe.
    Ⓑ Wood is used to build houses and furniture.
    Ⓒ Many fuels come from plants.
    Ⓓ Plants have many uses.

Finding Facts ◆ **Choose the word that best answers each question. Fill in the circle for your answer.**

5. Which part of a plant makes food?

    Ⓐ roots
    Ⓑ stem
    Ⓒ leaves
    Ⓓ flower

6. Which part of a plant protects its seeds?

    Ⓐ roots
    Ⓑ fruit
    Ⓒ flower
    Ⓓ leaves

Check your answers on page 165.

Lesson 10

The Environment

environment
Everything that surrounds us, including air, water, soil, plants, and animals.

pollution
The dirtying or poisoning of the environment.

Of all the animals on earth, only people make big changes in the **environment**. We build cities, factories, and farms. These things help support a large human population, but they can also cause harm to the environment.

Water in lakes, ponds, streams, and rivers often becomes polluted. Much of this **pollution** comes from chemical fertilizers used to grow crops. Some fertilizer is washed out of the soil by rain, and it flows into nearby lakes and streams. Fertilizer increases the growth of algae on the surface of the water. This blanket of algae blocks sunlight from reaching plants below. These plants die and decay, using up the oxygen in the water. Then fish start to die because there isn't enough oxygen. Eventually, most plant and animal life in the pond or lake dies.

Cities, suburbs, and factories add to water pollution, too. In some places, wastes from sewers run into streams and lakes. When it rains, water runs off highways and parking lots. This sometimes sends spilled oil and other chemicals into lakes and streams. Factories sometimes spill wastes into lakes and streams. These wastes can poison animals that live in the water.

Water pollution is not a new problem. In 1965, it was such a serious problem that the United States passed the Water Quality Act. This law was passed to keep wastes and chemicals out of our water. Although water pollution is still a problem, many lakes and rivers are cleaner today than they were in the 1960s.

DDT
A chemical that kills insects.

food chain
A cycle in which plants are eaten by animals, who are eaten by other animals.

The environment can also be damaged in ways that are harder to see than water pollution. For example, chemicals can get into the food supply. A well-known example is **DDT**, a chemical that kills harmful insects. Unfortunately, DDT also killed useful insects, like ladybugs. Over time people began to notice that DDT was killing other animals. Birds, fish, frogs, and other animals that eat insects had DDT in their bodies. Other animals in the **food chain** had DDT in their bodies. Farm animals, close to DDT spraying, also had DDT in their bodies. Traces of DDT were found in our meat and milk. Because DDT was so harmful, it was outlawed in 1972.

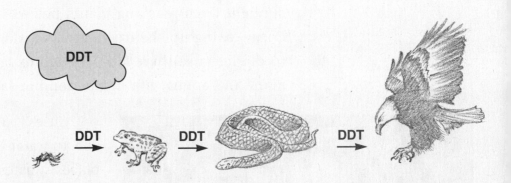

bovine growth hormone (bGH)
A substance that makes cows produce milk.

When a new chemical is used, it is not always clear whether it will later be found unsafe. For example, recently there have been arguments over **bovine growth hormone (bGH)**. BGH is a hormone that cows have naturally. When cows are given extra bGH, they make 10 to 20 percent more milk. Supporters of bGH say the milk is just the same as the milk from untreated cows. Others argue against giving cows bGH. They say that the cows get more infections of the udder. When the cows are given antibiotics to cure the infections, these drugs end up in the milk we drink. In this case, the U.S. government approved the use of bGH. It was decided that the milk was safe.

Lesson 10

Practice

Vocabulary in Context ◆ **Write the word that best completes each sentence.**

1. Humans are the only animals to make big changes in their _____.

2. The dirtying or poisoning of our surroundings is called _____.

3. _____ is a chemical that kills insects.

bGH
DDT
environment
pollution

Finding Facts ◆ **Choose the word or words that best answer each question. Fill in the circle for your answer.**

4. Which of the following is a large source of water pollution?
   Ⓐ chemical fertilizers
   Ⓑ farm crops
   Ⓒ rainwater
   Ⓓ bGH

5. Why would farmers want to give their cows extra bGH?
   Ⓐ to reduce the use of antibiotics
   Ⓑ to reduce the chance of getting udder infections
   Ⓒ to make the milk safe to drink
   Ⓓ to increase the amount of milk

Reading a Diagram ◆ **Look at the diagram on page 54. Write your answers below.**

6. How did DDT enter the food chain?

   _____

   _____

7. What is the main idea of the diagram?

   _____

   _____

Check your answers on page 165.

Strategies for Success

Classifying

To better understand nature, scientists place things into groups. Putting similar things into groups is called classifying. All the items in a group have something in common. A simple example of classifying is to group living things into plants and animals.

Scientists classify plants into several groups. Each group can be divided into several smaller groups.

SEED PLANTS: green plants that have roots, stems, leaves, and seeds

EVERGREENS: seed plants that have seeds protected by a cone

FLOWERING PLANTS: seed plants that have seeds protected by a fruit

❖ **STRATEGY: Read the material. Then think about similarities and differences.**

1. Look for key words like *group, divide, classify, similar, different, categories.*

2. Look for lists. Sometimes items that have something in common are grouped or classified in lists.

3. If you are asked to classify something, ask yourself: How is this similar to other things I might group it with? How is it different?

Exercise 1: Look again at the classification of seed plants. Which kind of a plant produced each of the following?

1. apple _____

2. pinecone _____

3. rose _____

Exercise 2: Read the paragraph below. Then classify each animal. Write the name of each animal group.

Animals are classified according to what they eat. Herbivores eat only plants, and carnivores eat only meat. Omnivores are animals that eat both meat and plants.

1. Mountain lions eat deer, as well as rabbits and other small animals.

   _____

2. Raccoons eat fish, mice, and berries.

   _____

3. Giraffes eat leaves and bushes. They also eat grasses.

   _____

Exercise 3: How would you classify the following living things? Put the items into groups and give each group a name. (Note: There is more than one way to classify them.)

    cat   bluebird   rose   tulip

    mosquito   dog   fly   eagle

Name of group ___things with fur_____

Members ___cat, dog_____

Name of group _____

Members _____

Name of group _____

Members _____

Name of group _____

Members _____

Check your answers on page 165.

Thinking and Writing

1. Three lessons in this unit talk about the prevention and cure of disease. What do antibodies, the skin, and vaccination have in common?

   _____
   _____
   _____

2. If you could develop a cure for any disease, which one would you cure? Why?

   _____
   _____
   _____
   _____

3. Three lessons in this unit discussed ways that people affect the environment. What do the loss of the Great Plains, endangered species, and pollution have in common?

   _____
   _____
   _____
   _____

4. If you could do one thing to protect the environment, what would you do? Why?

   _____
   _____
   _____
   _____

Check your answers on page 165.

Unit 2
EARTH SCIENCE

What do you already know about earth science? **Write something that you know about it.**

_____

_____

Preview the unit by looking at the titles and pictures. **Write something that you predict you will learn about earth science.**

_____

_____

Earth science is the study of the earth—how it formed and how it changes.

In this unit you will learn about:
- the air and ocean
- earthquakes and volcanoes
- hot spots and hurricanes
- pollution and recycling

Lesson 11

The Atmosphere

solar system
Our sun and its nine planets.

atmosphere
All the air surrounding the earth.

water vapor
The moisture in the air.

260°F, −280°F
A temperature of 260 degrees Fahrenheit, a temperature of minus 280 degrees Fahrenheit.

As far as we know, none of the other planets in our **solar system** can support life. There is a wonderful life-giving difference between our earth and other planets. Earth, unlike other planets, is surrounded by a blanket of air called the **atmosphere**. This atmosphere is a mixture of invisible gases, mostly nitrogen, oxygen, and **water vapor**. We call this mixture *air*. Without air, our planet would be without life.

Our moon is without life because the surface of the moon has no blanket of air to protect it. On the moon it is **260°F** during the day and **−280°F** at night. Life as we know it isn't possible under those extreme conditions.

In contrast, the atmosphere helps keep the earth's temperature even. The atmosphere extends thousands of miles from the earth, but it isn't the same throughout. The farther from earth you go, the thinner the air is. Nearly three-fourths of all the air in the atmosphere is within six miles of the earth's surface.

The Four Layers of the Earth's Atmosphere

Thermosphere

Mesosphere

Stratosphere (Ozone layer)

Troposphere

Until recently, people seldom worried about the atmosphere. It seemed endless and indestructible. Today we know better. Scientists have studied and learned a lot about the atmosphere and how it has been affected by natural disasters on earth. Read about the following examples.

Unit 2

volcano
An opening in the earth's crust through which melted rock is forced.

erupt
Explode violently.

cubic mile
An area one mile long, one mile wide, and one mile high.

Volcanic Eruptions
- In November 1985, the Colombian **volcano**, Nevado del Ruiz, **erupted**. The explosion caused millions of tons of gas and ash to enter the atmosphere. Temperatures nearby fell 20 degrees.
- In August 1883, Krakatoa, a volcano near Java, erupted. It sent nearly five **cubic miles** of rock and ash into the air. Darkness lasted two and a half days as far as 50 miles from the volcano. Temperatures fell. Fine dust rose into the atmosphere and traveled around the earth.

Why did temperatures fall when the air was filled with ash and dust? The temperatures fell because the sun's light was blocked by dust. Volcanic dust causes another problem. The dust can rise high in the air. If it rises high enough, the dust will ride the winds and circle the earth for months or even years. This dust blocks sunlight.

A drop in temperature of a few degrees may not seem like much, but it is important. Dinosaurs once ruled the earth for 100 million years. Many scientists believe that the dinosaurs died because temperatures around the word fell a few degrees. Over millions of years, plants and animals get used to their climate. When the earth's temperature changes even a few degrees, millions, perhaps billions, of plants and animals die.

Dinosaurs ruled the earth for 100 million years.

Lesson 11

nuclear winter
A severe drop in temperature resulting from a nuclear explosion.

greenhouse effect
The warming of the earth caused by an increase of carbon dioxide in the atmosphere.

carbon dioxide
A colorless, odorless gas produced when fuel is burned.

There is a close and very delicate relationship between the atmosphere and world climate. People have a large impact on the atmosphere. Because of this, scientists are concerned about a number of problems people could cause. Among the problems that worry scientists is a possible **nuclear winter**. In a nuclear winter, the dust and smoke from a nuclear explosion would block out the sun's rays. Temperatures would drop. Crops and animals would freeze. People would starve.

Another problem that worries scientists is the **greenhouse effect**. In a greenhouse, the heat from the sun gets trapped by the glass. This heat keeps everything in the greenhouse warm. Because of the increased amount of coal and oil we are burning, more **carbon dioxide** (CO_2) is present in the atmosphere than in past years. This carbon dioxide acts like glass in a greenhouse. It lets the sun's rays in, but it won't let much heat out. Many scientists predict that in another 100 years the earth will be warmer than it is now.

Many people are trying to decrease the amount of coal and oil they burn. This may help reduce the amount of warming. If the earth gets warmer, the climate may change in some areas. When the climate gets warmer, more water stays in the atmosphere instead of falling as rain. This will be good in some places. For example, areas that are now too rainy or too cool for good farming will be able to grow more crops. Other areas, however, will have more dry spells and will grow fewer crops.

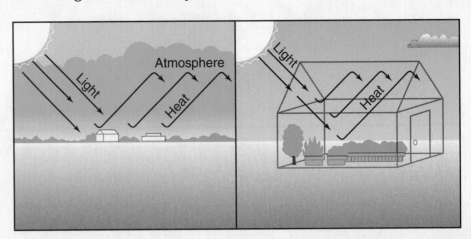

The greenhouse effect: Carbon dioxide in the atmosphere acts like glass in a greenhouse.

Unit 2

Lesson 11

Practice

Vocabulary in Context ◆ **Write the word that best completes each sentence.**

1. All the air surrounding the earth is called the _____ .

2. The drop in temperature caused by the dust and smoke from an nuclear explosion is called a nuclear _____ .

3. The gases in the atmosphere let sunlight in but do not let all the heat out. This is known as the _____ effect.

atmosphere
dioxide
greenhouse
winter

Finding Facts ◆ **Choose the words that complete each sentence. Fill in the circle for your answer.**

4. The earth's atmosphere
   Ⓐ causes the extinction of millions of plants and animals.
   Ⓑ helps keep the earth's temperature even.
   Ⓒ keeps sunlight from reaching the earth's surface.
   Ⓓ extends six miles from the earth.

5. An increase in the greenhouse effect has caused an increase in the amount of
   Ⓐ oxygen in the atmosphere.
   Ⓑ nitrogen in the atmosphere.
   Ⓒ carbon dioxide in the atmosphere.
   Ⓓ water vapor in the atmosphere.

Reading a Diagram ◆ **Write your answer below.**

6. Use the diagram on page 62 to describe how carbon dioxide gas and greenhouse glass are similar.

_____

_____

_____

Check your answers on page 166.

Lesson 12

Ocean Currents

At 2:20 A.M. on April 15, 1912, the *Titanic* sank. It was the largest and most luxurious ocean liner that had ever been built. The ship took 1,552 people down with it into the icy sea.

Earlier that evening, at 11:40 P.M., a group of young men were playing cards. They heard a grinding sound, so they went out on deck. There they saw a white mountain of ice. The *Titanic* had struck an **iceberg**.

Everyone thought that the ship was indestructible and couldn't sink. The men went back to their card game. The *Titanic* sailed on into the night, but not for long. At midnight, the captain admitted that the "unsinkable" ship was, in fact, going to sink.

What was a giant iceberg doing in the Atlantic Ocean 2,000 miles from where it was formed? It was because the Labrador Current brought it there.

iceberg
A very large piece of ice floating in the sea.

Icebergs are carried by ocean currents.

Unit 2

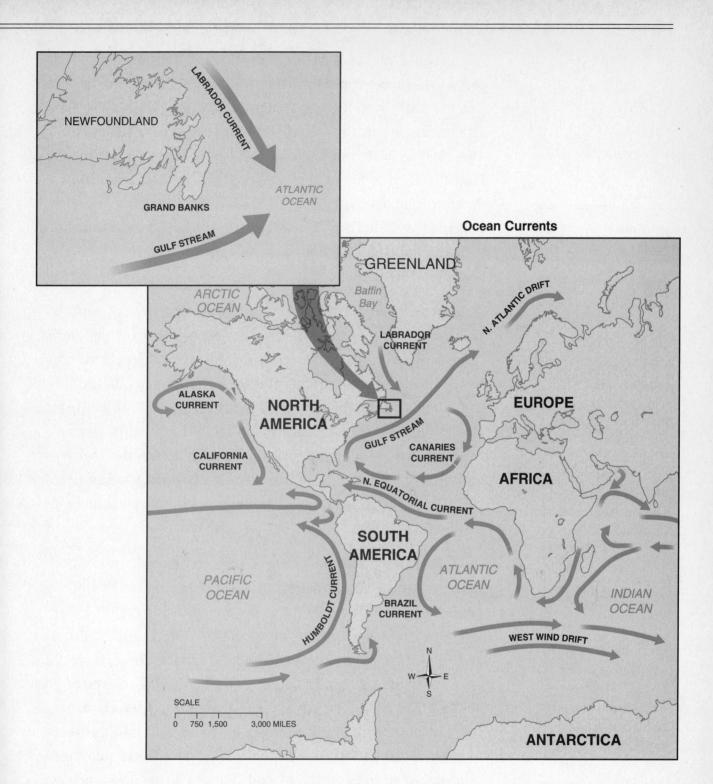

Ocean Currents

current
A movement of water in the ocean. It's like a river.

The water in the oceans is always moving. The water circulates by means of ocean **currents**. The ocean currents carry more water than any river on earth. The Gulf Stream, one of the most famous ocean currents, carries 1,000 times more water than the Mississippi River, the largest river in the United States.

Lesson 12

Because of the large amount of water that ocean currents carry, they have a major effect on climate. A warm current like the Gulf Stream keeps northern Europe warm. The Gulf Stream begins in the warm waters of the Gulf of Mexico. The current runs north along the East Coast of the United States and then heads out across the Atlantic Ocean to northern Europe. A cold current like the Labrador Current **originates** in Baffin Bay above the Arctic Circle. From there it heads south along the coast to Labrador and on to the coast of New England.

Ocean currents will carry anything that floats—bottles, ships, rafts, and even icebergs. The iceberg that hit the *Titanic* came from Greenland. The iceberg formed when a mountain of ice broke off one of the **glaciers** that cover much of Greenland. The iceberg floated south in the Labrador Current along the coast of Labrador until it hit the warm waters of the Gulf Stream where it began to melt. Before the iceberg melted, it crossed one of the busiest **shipping lanes** in the world. There it hit the *Titanic* and tore a 300-foot hole in the ship's side.

The same Labrador Current that can carry dangerous icebergs to the major North Atlantic shipping lanes also has good features. This current is responsible for creating one of the world's greatest fishing grounds. The Labrador Current and the Gulf Stream meet at the Grand Banks. The Grand Banks are rich fishing grounds because the mixing of the warm and cold waters produces perfect conditions for **plankton**. Since fish eat plankton, many fish gather here. The fishing fleets of many nations fish the Grand Banks. The fleets take home millions of dollars worth of cod, haddock, herring, and mackerel every year—due to the ocean's currents.

---

originate
Begin.

glacier
A mass of ice that flows slowly over land.

shipping lane
The official route that ships take between one port and another.

plankton
Microscopic plants and animals that live in the ocean.

Lesson 12

Practice

Vocabulary in Context ◆ **Write the word that best completes each sentence.**

1. A river of water in the ocean is an ocean _____ .

2. A flowing mass of ice over land is a _____ .

3. An _____ is a huge chunk of ice floating in the ocean.

current
glacier
iceberg
plankton

Finding the Main Idea ◆ **Choose the words that best answer each question. Fill in the circle for your answer.**

4. What is the main idea of the article on pages 64–66?

    Ⓐ The *Titanic* hit an iceberg and sank.
    Ⓑ Ocean currents have important effects on our lives.
    Ⓒ Icebergs can float 2,000 miles from where they were formed.
    Ⓓ The Gulf Stream is a famous ocean current.

5. What is the main idea of the last paragraph on page 66?

    Ⓐ Fish eat plankton.
    Ⓑ The fishing fleets of many nations fish the Grand Banks.
    Ⓒ The Labrador Current carries dangerous icebergs.
    Ⓓ The Labrador Current is responsible for the rich Grand Banks fishing grounds.

Finding Facts ◆ **Choose the words that complete each sentence. Fill in the circle for your answer.**

6. The Labrador Current is

    Ⓐ the name of a ship.
    Ⓑ a cold ocean current.
    Ⓒ a glacier in Greenland.
    Ⓓ the ocean current that warms Europe.

7. The Gulf Stream brings warm water to

    Ⓐ southern California.
    Ⓑ the Gulf of Mexico.
    Ⓒ northern Europe.
    Ⓓ South America.

Check your answers on page 166.

Lesson 13

Volcanoes

volcano
An opening in the earth's crust through which melted rock is forced.

lava
Melted rock.

eruption
An explosion of ash, steam, and lava from a volcano.

When explorer William Clark saw Mount St. Helens in 1805, he called it "perhaps the greatest" mountain in America. It was about 10,000 feet tall and had a perfect cone shape. The top was covered with snow. The neighboring American Indians, however, knew it was a **volcano**. They called it Tah-one-lat-clah, which means "fire mountain."

Between 1832 and 1857, Mount St. Helens gave off ash, steam, and **lava**. These were not serious **eruptions**, though. Over the years, people in the state of Washington got used to their quiet, beautiful mountain.

During the spring of 1980, the volcano rumbled and gave off some ash. Then on May 18, Mount St. Helens erupted with great force. One side of the mountain blew away. Rock, steam, ash, and lava came shooting out, killing 57 people. The explosion knocked down trees in a 232-square mile area. For 12 hours, ash fell in the northwestern U.S. and Canada.

Mount St. Helens erupted on May 18, 1980.

crater
A funnel- or bowl-shaped pit at the top of a volcano.

The perfect cone of Mount St. Helens is gone. In its place, 1,313 feet lower, is a rough, gray **crater**. Fallen dead trees that look like match sticks cover large areas. Plants and animals started returning to the land soon after the eruption. Still, the U.S. Forest Service estimates it will take 100 years for the land to recover.

This is what the area around Mount St. Helens looked like after the eruption.

What makes a volcano that had been quiet for almost 150 years suddenly explode?

crust
A hard outer covering.

plate
A section of the earth's crust.

molten
Melted.

The earth's **crust** is divided into a number of regions called **plates**. These plates are between 20 and 150 miles thick. They float on a bed of **molten** rock. These plates move slowly but surely at a speed of a few inches a year. Where two plates meet going in opposite directions, there is a head-on collision. Where one plate rides up over another plate, volcanoes are found.

The eruption of Mount St. Helens is a result of plate movement. The continent of North America has been moving west for millions of years. As it moves west, it rides up and over the Juan de Fuca plate of the nearby Pacific Ocean. As this happens, the Juan de Fuca plate is forced down under

Lesson 13

core
The center.

North America. As the plate is forced lower and lower toward the earth's molten **core**, the edge of the plate begins to melt. Melted rock starts to bubble up toward the surface. The result is a volcanic eruption.

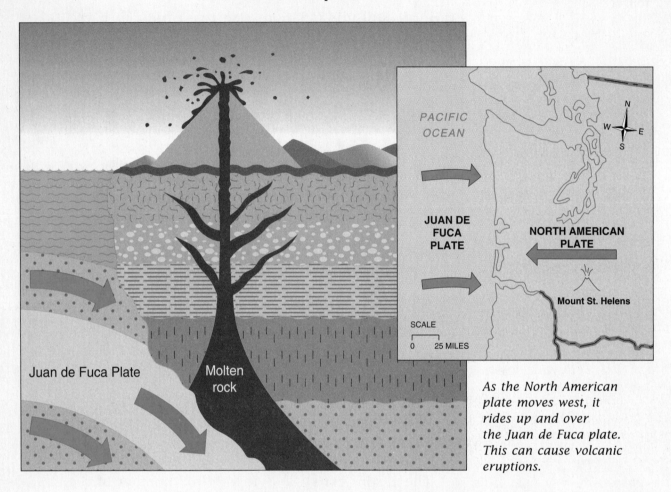

As the North American plate moves west, it rides up and over the Juan de Fuca plate. This can cause volcanic eruptions.

Will Mount St. Helens erupt again? Probably. The earth's plates are always moving. It isn't likely that they are going to change their speed or their direction. It is really just a matter of time until Mount St. Helens erupts again. The more we know about a volcano's past, the more we can tell about its future. If Mount St. Helens repeats its pattern, it will have another major eruption in less than 1,000 years.

Lesson 13

Practice

Vocabulary in Context ◆ **Write the word that best completes each sentence.**

1. A volcano's violent explosion is called an _____ .

2. A _____ is a bowl-shaped pit at the top of a volcano.

3. A _____ is a large, slowly moving section of the earth's crust.

crater
eruption
plate
volcano

Finding Facts ◆ **Choose the words that best answer each question. Fill in the circle for your answer.**

4. What events in 1980 gave warning that Mount St. Helens might erupt soon?

   Ⓐ Wildlife left the area.
   Ⓑ The weather became much colder than usual.
   Ⓒ The volcano rumbled and gave off ash.
   Ⓓ Lava flowed down the mountainside.

5. What is at the top of Mount St. Helens today?

   Ⓐ a perfect cone
   Ⓑ an evergreen forest
   Ⓒ flowing lava
   Ⓓ a rough, gray crater

6. Which of the following usually happens when a volcano erupts?

   Ⓐ Lava comes out.
   Ⓑ Plates change their speed.
   Ⓒ Snow falls.
   Ⓓ both A and B

7. What is true about the plate on which Mt. St. Helens occurs?

   Ⓐ It is called the North American plate.
   Ⓑ It rides up and over the Juan de Fuca plate.
   Ⓒ It moves in the opposite direction of the Juan de Fuca plate.
   Ⓓ all of the above

Check your answers on pages 166–167.

Strategies for SUCCESS

Reading a Map

A map is a diagram that shows where places are. Sometimes maps also show features of a place. Here is a map of the United States. It also shows the weather on one day. This map is like the weather maps you see in the newspaper.

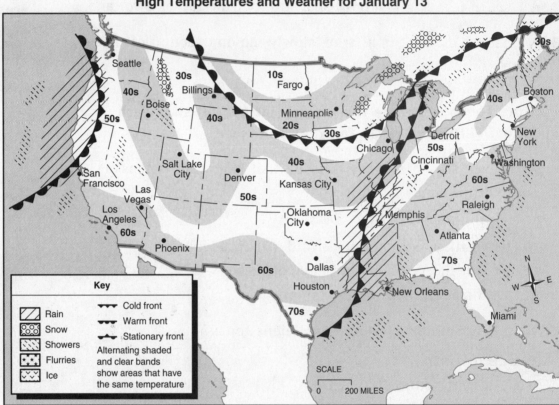

When you look at a map, first figure out the main idea. The main idea is often stated in the map's title. The main idea of this map is the temperature and weather on January 13.

❖ **STRATEGY: Look at the map. Ask yourself: What is the main idea?**

1. Look at the title or caption. It tells you the main idea of the map.

2. Look at the whole map. Ask yourself what this map shows.

Unit 2

Many maps have clues to help you find information:
- The long arm of the compass rose always points north. Look above Miami, Florida, for the compass rose.
- The scale shows how to measure distance. Look below Houston for the scale. For example, the distance between Dallas and Houston is about 250 miles.
- The map key explains how things are shown. Look on the lower left for the key. The map key symbols show different types of weather—rain and snow. It also shows where the **fronts** are.

front
Boundary between types of weather.

❖ **STRATEGY:** **Look at the map. Ask yourself: What detailed information does this map show?**

1. Look for the compass rose. It points north.
2. Look for the scale. It helps you measure distance between places on the map.
3. Look for the key. It explains how things are shown.

Exercise: **Use the weather map to answer these questions. Write your answers below.**

1. Which city is directly north of Kansas City?
   _____

2. Which city is almost directly south of Dallas?
   _____

3. Which is closer to New York—Boston or Raleigh?
   _____

4. How many miles apart are Atlanta and Washington DC?
   _____

5. What is the temperature in Atlanta? _____

Check your answers on page 167.

Lesson 14

Earthquakes

At 4:31 A.M. on January 17, 1994, people in Northridge, California were awakened by the loud rumbling and shaking of an earthquake. For 40 seconds, the Los Angeles suburb shook. An apartment complex collapsed, killing 16 people. Fifteen miles away, part of Interstate Highway 10 fell down. All over the city, buildings swayed and things came crashing down. Some buildings fell, and others were damaged so badly that people could not go back into them. Oil and gas lines broke, causing explosions. Tens of thousands of homes lost water or electricity. Other major highways collapsed. Those 40 seconds caused over 15 billion dollars of damage. All told, the Northridge earthquake killed 55 people and injured 4,000.

Despite the death toll and damage, the Northridge earthquake was medium sized. It was not "The Big One." The Big One is the major earthquake that scientists predict will occur in California. However, the Northridge earthquake started in the rocks directly below the city of Northridge. In a city, there are many people and buildings close together. That's why the Northridge earthquake caused so much damage, as did earlier quakes in San Francisco in 1906 and the San Fernando Valley in 1971.

The Northridge earthquake did great damage.

Unit 2

Earthquakes are caused by movements of the earth's crust. The same process that causes volcanic eruptions and mountain formations causes earthquakes. The earth's crust is not one piece. Instead, the earth is made up of about a dozen major plates and a number of smaller plates. These plates cover the earth like a jigsaw puzzle. Yet unlike a jigsaw puzzle, the plates are not locked together. Instead, the plates float on the molten **mantle** beneath them.

mantle
The part of the earth between the crust and the core.

However, the different plates that form the earth's crust don't float around in the same way. For reasons we do not yet understand, each plate moves at a speed and direction of its own. The Pacific plate under the Pacific Ocean is moving to the northwest. The North American plate is moving west. As the Pacific and North American plates slide past each other, they sometimes get "stuck." When this happens, the pressure builds up until there is an earthquake. Then the plates loosen and continue on their way.

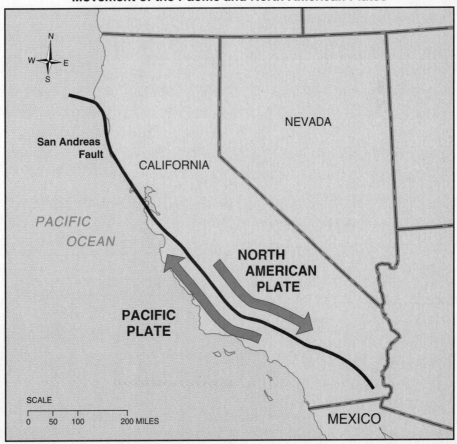

Movement of the Pacific and North American Plates

Lesson 14

fault
A crack in rock along which movement occurs.

Is it possible to predict where and when an earthquake will occur? Predicting "where" is not very difficult. In California, you can actually see where the two plates meet along the San Andreas **fault**. The fault runs more than 800 miles, from Mexico to Cape Mendocino. Earthquakes in California usually occur on or near the San Andreas fault. The Northridge earthquake of 1994 was west of the San Andreas in an area of smaller faults.

Predicting when an earthquake will occur is more difficult. Along the San Andreas fault, earthquakes happen when the plates very suddenly start sliding again after being stuck together. Scientists have located areas along the fault that have been stuck for more than a century. Scientists know pressure is building in these places. However, they don't know how much pressure there is or when it will be released. Thus, "The Big One" could happen in any of these places where pressure is building at any time.

The San Andreas fault can be seen for hundreds of miles.

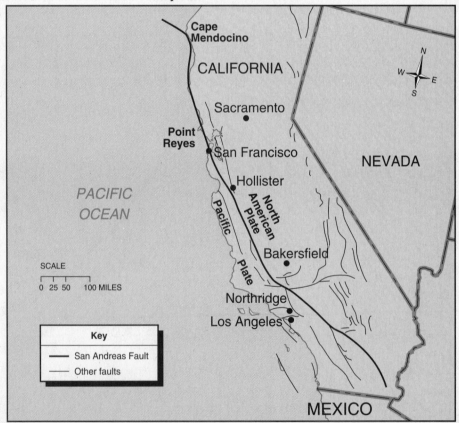

Unit 2
76

Lesson 14

Practice

Vocabulary in Context ◆ Write the word that best completes each sentence.

1. The part of the earth between the crust and the core is the _____ .

2. A crack in rock along which movement occurs is a _____ .

3. A violent shaking of the earth is an _____ .

| earthquake |
| fault |
| jigsaw |
| mantle |

Finding the Main Idea ◆ Choose the words that best answer each question. Fill in the circle for your answer.

4. What is the main idea of the first paragraph on page 76?

   Ⓐ Earthquakes in California usually occur near the San Andreas fault.
   Ⓑ Predicting where an earthquake will occur is not very difficult.
   Ⓒ The Northridge earthquake was west of the San Andreas fault.
   Ⓓ Predicting when an earthquake will occur is not very difficult.

5. What is the main idea of the last paragraph on page 76?

   Ⓐ Earthquakes happen when the plates very suddenly start sliding again after being stuck together.
   Ⓑ "The Big One" could happen at any time.
   Ⓒ Predicting when an earthquake will occur is difficult.
   Ⓓ Some areas along the fault have been stuck for more than a century.

Reading a Map ◆ Look at the map of California on page 76 to answer the questions. Write your answers below.

6. Where are most of the faults located?

   _____

7. What are the names of two cities located on or near a fault?

   _____

Check your answers on page 167.

Lesson 15

Hot Spots

The Hawaiian Islands are a chain of volcanic islands in the middle of the Pacific Ocean. They are 1,200 miles away from the coast of California.

The Hawaiian islands were formed in the middle of the Pacific plate over a **hot spot**. A hot spot is like a **blowtorch**. The hot spot heats the crust above it so that the crust expands and bubbles up. The result is a kind of volcanic eruption.

Right now the island of Hawaii, the largest of the Hawaiian Islands, sits on top of a hot spot in the middle of the Pacific Ocean floor. Its three active volcanoes continue to build the island. Two of them, Mauna Loa and Hualalai, are among the tallest mountains on earth. They rise more than 30,000 feet above the sea floor.

hot spot
A very hot area of the earth's mantle right below the crust.

blowtorch
A tool that uses a mixture of gas and air under pressure. It makes a very hot flame.

The island of Hawaii sits on top of a hot spot.

Unit 2

However, as the Pacific plate slowly moves northwest, Hawaii will move off the hot spot and a new volcanic island will rise from the ocean floor. This is the way all of the Hawaiian Islands have been formed.

Nobody knows what causes these hot spots under the earth's crust. There are **theories** but no facts yet. Scientists have located at least 30 hot spots around the world. They are responsible for creating many islands in the ocean. Bermuda, off the eastern coast of the United States, may have been built by a hot spot that is now extinct.

theory
An idea or set of ideas that tries to explain why certain things happen.

geyser
A natural fountain of hot water and steam that rises suddenly into the air.

fumarole
A hole in the ground where smoke and gases escape.

spring
A place where water rises out of the ground.

A hot spot heats the earth's crust above it. The result is a volcanic eruption.

Wherever there is a hot spot, there will be volcanic activity. But not all hot spots create islands. Sometimes a hot spot is located under a continent. A hot spot under a continent causes **geysers**, **fumaroles**, and hot **springs**.

Lesson 15

Yellowstone National Park in Wyoming is located over a hot spot. Yellowstone's volcanic activity makes it one of the most popular national parks. The park includes 200 geysers, 10,000 hot springs, and many fumaroles. The most famous geyser at the park is called Old Faithful. Old Faithful was named because it has sent a column of hot water and steam 150 feet in the air about once an hour for the last 100 years.

About 600,000 years ago, there was a tremendous volcanic explosion where Yellowstone National Park is now. That explosion was 1,000 times more powerful than the recent eruption of Mount St. Helens in Washington state. Today, volcanic explosions may not be happening in Yellowstone, but steam is certainly exploding from geysers. Scientists know that geysers only form where underground water is being heated by very hot crust. As a result, they know the Yellowstone hot spot is still hot indeed.

In Yellowstone National Park, Old Faithful sends a column of hot water and steam into the air about once an hour.

Lesson 15

Practice

Vocabulary in Context ◆ Write the word that best completes each sentence.

1. A very hot area of the mantle below the crust is a _____ .

2. A natural fountain of hot water and steam that rises suddenly out of the ground is a _____ .

| fumarole |
| geyser |
| hot spot |

Finding the Main Idea ◆ Choose the words that best answer the question. Fill in the circle for your answer.

3. What is the main idea of the last paragraph on page 80?
   Ⓐ There was a major eruption at Yellowstone about 600,000 years ago.
   Ⓑ That explosion was 1,000 times more powerful than the eruption of Mount St. Helens.
   Ⓒ Volcanic explosions are not happening in Yellowstone today.
   Ⓓ The fact that Yellowstone has geysers suggests that the area is still an active hot spot.

Finding Facts ◆ Choose the word or words that best complete each sentence. Fill in the circle for your answer.

4. Volcanoes that form in the middle of a plate result from
   Ⓐ fumaroles.
   Ⓑ one plate colliding with another.
   Ⓒ a hot spot in the earth's mantle.
   Ⓓ plates moving apart.

5. Yellowstone National Park has interesting
   Ⓐ island scenery.
   Ⓑ geysers, hot springs, and fumaroles.
   Ⓒ active volcanoes.
   Ⓓ glaciers.

Check your answers on page 167.

Lesson 16

Hurricanes

hurricane
A violent tropical storm with winds of more than 75 miles per hour.

meteorologist
A scientist who studies and predicts the weather.

Most people who live in the Caribbean, around the Gulf of Mexico, or on the Atlantic coast of the United States have probably lived through a **hurricane**. One of the worst or most terrifying hurricanes was Hurricane Andrew in 1992. In that hurricane 65 lives were lost, and billions of dollars of damage was done in Florida and Louisiana.

Meteorologists give hurricanes names. The first hurricane of the season gets a name that starts with the letter *A*. The second hurricane gets a name that starts with the letter *B* and so on. For many years, hurricanes were given only women's names. Now both men's names and women's names are used.

What is a hurricane like? First comes the rain. Gradually, the rain gets heavier, and the winds get worse. To be called a hurricane, a storm must have winds of at least 75 miles per hour. Some hurricanes, like Andrew, have winds of 180 miles per hour. As the winds get stronger, a hurricane begins to destroy things in its path. Trees are blown over and crash down on houses, cars, and telephone poles. Roofs and walls are blown off buildings. Electricity and telephone services go out.

Hurricanes cause millions of dollars in damage.

The rain continues. Up to ten inches of rain may fall in a day. The tides are much higher than normal. When Hurricane Camille came ashore in Louisiana in 1969, the tides in the Gulf of Mexico were 25 feet above normal. The floods that resulted killed 250 people.

What causes a hurricane? Hurricanes form over **tropical** seas when the water temperature is over 80°F. Hurricanes occur in the summer or early fall when the oceans are their warmest. The warm, moist air rises. As it rises, it starts to swirl in a **counter-clockwise** direction. It's like pulling the plug out of the sink and watching the water swirl faster and faster down the drain. The rising air swirls faster. As it rises, this warm air cools. As it cools, rain starts to fall because cool air cannot hold as much moisture as warm air.

tropical
Very warm or hot, and usually near the equator.

counter-clockwise
Moving in a circle from right to left.

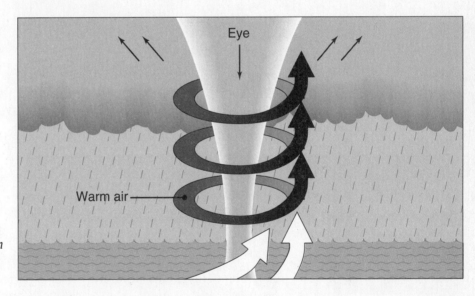

Diagram of a hurricane: warm air rises in a counter-clockwise direction. As the warm air rises, it cools, and rain starts to fall.

eye
The calm area in the middle of a storm.

If the hurricane stays over warm seas, it may grow until the **eye** of the hurricane is 60 miles across and the swirling winds are 300 miles from the center of the eye.

Hurricanes do not always stay out in the ocean where they form. They move at an average speed of 16 miles per hour. When they move over land, the rain continues, but the winds slow down. Meteorologists are getting better at predicting where a hurricane is headed and how quickly it will arrive.

Lesson 16

Not long ago, hurricanes struck without warning. Many lives were lost. Now weather satellites circle the earth frequently. They fly high above the clouds and take pictures of the clouds and the earth below. Because hurricane clouds look like a huge **pinwheel**, they are easy to recognize. Since 1966 when the first weather satellite was launched, no hurricane has ever taken us by surprise. In fact, we are usually warned of a coming hurricane many days in advance.

In recent years, **radar** has been used to find the path of hurricanes. Advance warning saves lives. Small boats can return to shore. People who live near the water can move to higher ground. Schools and offices can close. Without our modern warning systems, Hurricane Andrew would have taken many more lives than it did.

pinwheel
A toy made in the shape of a propeller attached to a stick.

radar
A device that uses radio waves to find and follow storms or other moving objects.

Satellite photograph of a hurricane. Notice the swirling (pinwheel) pattern of clouds.

◆ Extension
Stormy Weather

Although meteorologists can predict storms today so we can prepare for them, bad weather can still affect our lives. Tornadoes are funnel-shaped wind storms that can tear up trees and flatten houses. They are most common in the Midwest. Blizzards are very heavy snow storms. They affect the northern half of the country. Rainstorms, although not as windy as hurricanes, can also do damage. Heavy rains can cause rivers to flood nearby land. In steep areas, rainstorms can cause mudslides.

Describe a bad storm that you remember. How did it affect you or people you know?

_____

_____

_____

Check your answers on page 168.

Lesson 16

Practice

Vocabulary in Context ◆ **Write the word that best completes each sentence.**

1. A violent tropical storm with high winds is a _____ .

2. Weather satellites and _____ are used to find and follow a hurricane.

3. Hurricanes form over _____ seas when the water temperature is over 80°F.

> hurricane
> pinwheel
> radar
> tropical

Finding the Main Idea ◆ **Choose the words that best answer the question. Fill in the circle for your answer.**

4. What is the main idea of the first paragraph on page 84?

   Ⓐ Since 1966 when the first weather satellite was used, hurrricanes have not taken us by surprise.
   Ⓑ Weather satellites circle the earth frequently.
   Ⓒ Hurricane clouds look like a large pinwheel when seen from above.
   Ⓓ We are usually warned of a hurricane many days in advance.

Finding Facts ◆ **Choose the words that best complete each sentence. Fill in the circle for your answer.**

5. To be a hurricane, a storm's winds must blow at least

   Ⓐ 50 miles per hour.
   Ⓑ 75 miles per hour.
   Ⓒ 125 miles per hour.
   Ⓓ 180 miles per hour.

6. The eye of a hurricane is the

   Ⓐ part that is photographed from satellites.
   Ⓑ pinwheel-shaped wind pattern.
   Ⓒ leading edge.
   Ⓓ calm area in the middle.

Check your answers on pages 167–168.

Lesson 17

Pollution and Recycling

Each day, for every person in the United States, 3.6 pounds of garbage are thrown out. We throw out paper, plastics, metals, glass, food, and many other kinds of trash. In one year, we throw away 200 million tons (400,000,000,000 pounds!) of garbage. Where does it all go?

landfill
A place in the ground where garbage is buried.

Most of the garbage goes to **landfills**. When there were fewer people and more space, dumping garbage in landfills was not a problem. However, today we are running out of places to put garbage. In the Northeast, landfill space will run out in the next few years. In the rest of the country, 80 percent of the landfills will be full by then. Clearly, we must look for other ways to get rid of garbage.

pollution
The dirtying or poisoning of the environment.

There are several other ways to dispose of garbage. Garbage used to get dumped in the ocean. This has been stopped because it causes water **pollution**. Garbage has been burned, but burning can cause air pollution. A better solution is to make less waste in the first place by using simpler packaging. Packaging makes up about 33 percent of our garbage. We need to find ways to reuse our garbage instead of throwing it away.

recycling
Making garbage useful instead of just dumping it.

yard waste
Clippings from lawns and trees, raked leaves, and other plant garbage from yards.

Recycling helps decrease the amount of garbage that is dumped or burned. In the last few years, recycling programs have spread throughout the country. In 5,000 communities, people separate regular garbage from items to be recycled. Glass, plastic, metal, and newspapers usually can be recycled. To encourage people to recycle, the community picks up these materials at people's homes. In some communities there is even a separate collection for **yard waste**. To make sure that people sort their garbage, some cities like Seattle charge for each container of regular garbage it collects. Recycled items are collected free of charge.

Material to be recycled is brought to a recycling center. In communities that do not pick up materials for recycling, people can bring certain kinds of garbage to recycling centers themselves. There it is sorted further. The material we recycle is made into many new products. Some are very different from the original material.

This recycling plant in Rhode Island processes tons of aluminum, paper, glass, and plastic.

Lesson 17

fiber
Thin threads that make up fabric and other materials.

fiberfill
A material made from fibers that is used to create bulk or warmth.

insulation
A material used to hold in heat.

compost
A mixture of organic matter used to enrich soil.

This item is recyclable.

Many products made from recycled material have this symbol.

Where recycled materials go:
- Plastic soft-drink bottles are melted and made into pipes and bottles. The plastic can also be made into **fiber** for **fiberfill** for jackets, **insulation**, and synthetic rug yarn.
- Plastic milk containers are chopped, melted, and used to make toys, pails, and bottles for shampoo and detergent.
- Glass is sorted by color and then melted. It is formed into new bottles and jars.
- Aluminum cans are melted and made into new cans, cooking pots, lawn chairs, and siding.
- "Tin" cans, which are really steel, are melted. The steel is used for cans, tools, and some auto parts.
- Newspaper and other paper is mixed with water to make pulp. The pulp is used to make paper, cardboard, egg cartons, and building materials.
- Yard waste is used to make **compost**.

 Extension

Personal Recycling

Some people recycle things at home. Many people began recycling during World War II. At that time, metals, including aluminum foil and gum wrappers, were recycled for the war effort. After the war, some people continued to reuse things in their own homes. Plastic containers were used for food storage. Pieces of aluminum foil were reused.

Today, many people have the habit of using things once and then throwing them out. Even when something could be reused at home, it may end up in the trash.

What things do you use more than once before you throw them out? Write your ideas below.

_____

_____

Check your answers on page 168.

Lesson 17

Practice

Vocabulary in Context ◆ **Write the word that best completes each sentence.**

1. _____ is reusing garbage instead of throwing it away.

2. Yard waste can be used to make _____ .

3. A _____ is a place where garbage is buried.

compost
landfill
pollution
recycling

Finding Facts ◆ **Choose the words that best answer the question. Fill in the circle for your answer.**

4. Why are landfills filling up?

    Ⓐ More people are producing more garbage.
    Ⓑ Newspapers take a lot of space.
    Ⓒ Many people do not recycle their garbage.
    Ⓓ Metal containers can be reused.

5. How can a community make it easier for people to recycle?

    Ⓐ pass laws that make people sort their garbage
    Ⓑ raise taxes to pay for recycling centers
    Ⓒ collect materials for recycling at people's homes
    Ⓓ all of the above

Classifying ◆ **Choose the word or words that best answer each question. Fill in the circle for your answer.**

6. Which of the following is an example of pollution?

    Ⓐ reusing newspapers
    Ⓑ the dumping of garbage in the ocean
    Ⓒ making less garbage in the first place
    Ⓓ running out of places to put garbage

7. Which of the following is made from recycled materials?

    Ⓐ plastic foam cups
    Ⓑ wool yarn
    Ⓒ cardboard
    Ⓓ lawn clippings

Check your answers on page 168.

Strategies for SUCCESS

Using Context Clues

If you were reading and you came to a word you didn't understand, what would you do? You might look it up in a dictionary. However there is a faster way to try to find its meaning. Sometimes you can look at the other words in the sentence or passage and figure out what the word means. In other words, you can figure out the word from the context it's in. See if you can figure out the meaning of the word *gyring* in the following passage:

> If the hurricane stays over warm seas, it may grow until the eye—the calm area in the middle of the storm—is 60 miles across. The hurricane winds, in contrast, are gyring wildly around at distances up to 300 miles from the center.

The passage tells you that the area in the middle of the storm is calm. Whatever the winds are is in contrast to that calmness. *Wildly* also gives you a clue that *gyring* means "spinning" or "swirling."

> ❖**STRATEGY:** **Look for certain word clues.**
>
> 1. Find words that indicate similarities or differences.
>
> 2. Find words that describe the word that you want to know the meaning of.

Exercise 1: What do you think the word *spewed* means in the following passage? Write your answer below.

> In August 1883, Krakatoa, a volcano near Java, exploded. It spewed nearly five cubic miles of rock and ash violently into the air.

_____

_____

Krakatoa

Unit 2
90

Sometimes you have to read the entire paragraph surrounding the word in order to figure out what it means. Figure out what *curbside* means in the following paragraph.

Recently, curbside recycling programs have spread throughout the country. In 5,000 communities, people separate regular garbage and other items. Usually, glass, plastic, metal, and newspapers are packaged separately and collected for recycling. One truck may collect the regular garbage. A second truck collects the glass, metal, plastic, and paper. These items are taken to factories where they are cleaned and used again.

From this, you can guess that curbside recycling is a kind of garbage collection where trucks come to pick up certain types of garbage to be recycled.

❖ **STRATEGY: Think it through.**

1. Read and reread the words or sentences around the unknown word.
2. Use what you already know to figure out the meaning of the word.

Exercise 2: What do you think a *radiosonde* is, based on the following paragraph? Write your answer below.

The upper atmosphere is not an area that meteorologists can easily reach. That's why a radiosonde is so useful. Carried high above the earth by a helium-filled balloon, a radiosonde sends out radio signals. A special kind of radio back on the ground receives the signals. As a result a meteorologist can tell what the weather is like high above the clouds.

Meteorologist

Check your answers on page 168.

Thinking and Writing

1. Three lessons in this unit describe earthquakes and volcanoes. Earthquakes and some volcanoes, like Mount St. Helens, are caused by the same thing. What causes both earthquakes and volcanoes like Mount St. Helens?

   _____
   _____
   _____

2. A cycle is something that happens over and over again. Describe a cycle that happens on the earth and one that happens in the atmosphere.

   _____
   _____
   _____
   _____
   _____

3. What we do affects the earth and its atmosphere. Describe one way people are changing or have changed the earth that you think is a change for the better. Then describe a change caused by people that you don't think is a change for the better.

   _____
   _____
   _____
   _____
   _____

Check your answers on page 168.

Unit 3
CHEMISTRY

Chemistry is the study of matter. It deals with matter's properties, structure, and changes. If you have ever seen ice melt, baked a cake, held a balloon, or removed rust from a car or bicycle, you have experienced chemistry. What do you already know about chemistry?
Write something you know about it.

_____

_____

Preview the unit by looking at the titles and pictures. **Write something that you predict you will learn about chemistry.**

_____

_____

In this unit you will learn about:
- atomic structure
- states of matter
- conservation of matter

Lesson 18

Matter All Around

Stub your toe on a wooden chair, and you know just how hard wood can be! Dip your toe in a tub of warm water, and your toe will feel warm and wet. Wood and water are different, but they are both **matter**. Your toe, this book, rain, and even air all are made of matter.

matter
A substance that occupies space and has mass.

chemistry
The study of matter.

Matter is everywhere. It's no wonder that scientists are always trying to understand it by studying **chemistry**. Did you ever wonder why water forms on the outside of a glass of cold water? Do you know why wood and water feel so different? Chemists answer these questions by studying matter.

Air, water, flowers, wood, and sand are all forms of matter.

element
A substance that can't be broken down into simpler substances.

Matter is made up of many different substances called **elements**. You may already be familiar with some elements. Diamonds are made of the element carbon. Carbon is also what makes up the "lead" in your pencil. Carbon is found in all living things, including trees, leaves, and people. Gold and silver are elements that have been used for centuries to make jewelry and coins. Oxygen is an element found in air, water, crystals, glass, and sand.

atom
The smallest particles of an element.

nucleus
The center of an atom that contains protons and neutrons.

proton
A positively charged particle in an atom.

neutron
An uncharged particle in an atom.

electron
A negatively charged particle in an atom.

The smallest particles of an element with that element's features are called **atoms**. Scientists believe the structure of an atom looks something like our solar system. The solar system has a sun in the middle and planets that move around it. An atom has a **nucleus** in the middle and particles that move around it. The nucleus contains two kinds of particles—**protons** and **neutrons**. **Electrons** circle around the nucleus. They have very small masses. Protons and neutrons also have small masses, but they each have about 1,800 times the mass of an electron.

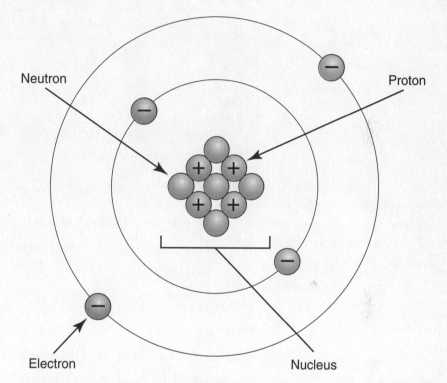

Atoms are made up of protons, neutrons, and electrons.

Electrons move around the nucleus at high speeds. They are found at different distances from the nucleus in what are called energy levels. The farther an electron is from the nucleus, the more energy it has.

All atoms of a certain element have the same number of protons and electrons. For example, all carbon atoms contain 6 protons and 6 electrons. All gold atoms have 79 protons and 79 electrons. No two elements have the same number of protons. Therefore, this number of protons—called an element's **atomic number**—is different for every element.

atomic number
The number of protons in an atom.

atomic structure
The number of protons, neutrons, and electrons that make up an atom.

The table below lists some common elements. Notice that these elements contain different numbers of protons and electrons. Different kinds of matter have different features. Their **atomic structure** is the reason. The atoms that make up the matter are not all the same. The atoms in a gold ring and in an aluminum can have different numbers of particles. They also have different atomic numbers.

Table of Elements			
Element	Symbol	Atomic Number (Protons)	Electrons
Hydrogen	H	1	1
Helium	He	2	2
Carbon	C	6	6
Oxygen	O	8	8
Neon	Ne	10	10
Sodium	Na	11	11
Aluminum	Al	13	13
Silicon	Si	14	14
Chlorine	Cl	17	17
Silver	Ag	47	47
Gold	Au	79	79

symbol
One or two letters that stand for an element.

Notice in the table that each element has a **symbol** made of different letters. These symbols make it easier to refer to the elements.

compound
A substance made of two or more elements.

Everything around you is made of atoms. Some materials are made up of only one kind of atom. Gold is made of only gold atoms because gold is an element. Water is made of both oxygen and hydrogen atoms, so it is a **compound**. You are probably familiar with many other compounds. Table salt is a compound made up of the elements sodium and chlorine. It is called sodium chloride. The table below lists some common compounds.

Compound	Elements
Salt	Sodium, Chlorine
Sugar	Carbon, Hydrogen, Oxygen
Glass	Silicon, Oxygen

Unit 3

Lesson 18

Practice

Vocabulary in Context ♦ **Choose the word or words that best complete each sentence.**

1. The _____ carbon is a common substance on our planet.

2. Sodium hydroxide is a _____ made up of the elements sodium, hydrogen, and oxygen.

3. Each element has a different _____ number.

4. Atoms contain _____ with a positive charge and _____ with a negative charge.

| atomic |
| compound |
| electrons |
| element |
| neutrons |
| protons |

Finding Facts ♦ **Choose the words that best complete each sentence. Fill in the circle for your answer.**

5. A compound is made up of

   Ⓐ atoms that all are alike.
   Ⓑ elements that all are alike.
   Ⓒ two or more elements.
   Ⓓ two or more atomic numbers.

6. The nucleus of an atom is made up of

   Ⓐ electrons and protons.
   Ⓑ protons and neutrons.
   Ⓒ neutrons and electrons.
   Ⓓ electrons, protons, and neutrons.

Using Context Clues ♦ **Choose the words that best complete the sentence. Fill in the circle for your answer.**

7. Based on the last paragraph on page 96, you can conclude that water is

   Ⓐ an atom.
   Ⓑ an element.
   Ⓒ a compound.
   Ⓓ a proton.

Check your answers on pages 168–169.

Lesson 19

States of Matter

If you take an ice cube out of the freezer and leave it on the kitchen table, it melts. It changes from a solid to a liquid. This is a change of **state**. If you put the water from the ice cube in a pan and heat it, the water will change its state again. It will become water vapor, a gas. Matter can change its state when there is a change in temperature. The three states of matter are **solid**, **liquid**, and **gas**.

Ice, water, and water vapor are the same compound but are in different states. They are all **H_2O**. Two atoms of hydrogen gas and one atom of oxygen become connected to form a **molecule** of water. Ice is H_2O, liquid water is H_2O, and water vapor is H_2O.

state
A condition of matter.

solid
The state of matter that has a volume that can't change and a shape that can't change.

liquid
The state of matter that has a volume that can't change but a shape that can change.

gas
The state of matter that has a volume that can change and a shape that can change.

H_2O
The chemical formula for water.

molecule
A stable group of two or more atoms.

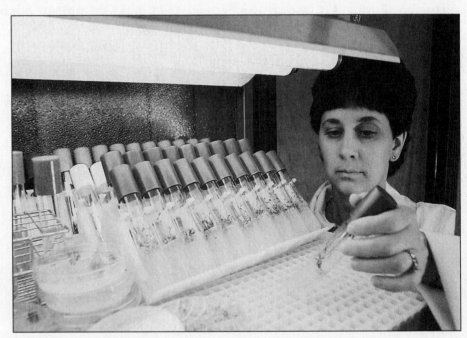

A scientist at work in a laboratory

What happens when H_2O changes its state? When water is in its solid state, ice, its molecules are lined up in an orderly pattern. The molecules of solid H_2O vibrate slightly back and

Unit 3
98

forth in their places. You must heat ice to change its state. The warmer the ice becomes, the faster the molecules vibrate. At some point during the heating, the molecules shake so fast that they don't stay in their places anymore. We call this movement melting. H_2O has changed from a solid state, ice, to a liquid state, water. It is still the same substance. Each molecule has exactly the same atoms in it.

The three states of matter: solid, liquid, and gas

Ice   Water   Steam

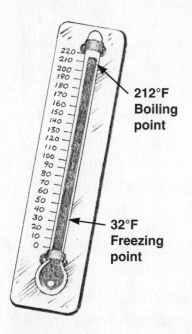

212°F Boiling point

32°F Freezing point

What happens as water in a pan is heated and becomes water vapor? As the water gets hotter, the molecules move faster and faster. They bounce off each other and off the sides of the pan. We say the water is boiling. Now the molecules near the surface of the water begin to bounce out into the air. They mix with the atoms in the air. At some point there are no water molecules left in the pan. All of the H_2O molecules still exist, but they moved into the air. The H_2O has changed from a liquid, water, to a gas, water vapor.

We see water change its state, from solid to liquid to gas, in our daily lives. Water freezes at 32°F and boils at 212°F. A normal kitchen refrigerator and stove will give us these temperatures. Most other substances do not change state at these temperatures. For example, you can't melt an iron frying pan on the stove.

Lesson 19

We think of most metals as solids because they are solids at the temperatures we live in. However, all metals melt if they're heated. Tin melts at 449°F. Iron must be heated to a much higher temperature, all the way up to 2,795°F, to melt. If a metal is heated enough, a liquid metal will boil away into a gas. Iron boils at a temperature of 5,432°F.

If you heat a solid, it will melt and then boil. This process can be reversed. Gases become liquids and then solids as they get colder. We think of oxygen as a gas because that is its state at room temperature. However, if we cool oxygen to −297°F, it becomes a pale blue liquid. If we cool it even further, it freezes into a solid at −361°F. The temperature at which a substance melts or freezes is the same. For oxygen, this temperature is −361°F.

The fact that different elements have different melting (or freezing) points is very important in our daily lives. The metal with the highest melting point of all is **tungsten**. It doesn't melt until it reaches a temperature of 6,116°F. This is why tungsten is used to make filaments in light bulbs.

Mercury, on the other hand, is a metal that is a liquid at room temperature. Mercury is sometimes used in thermometers to show the temperature. Mercury freezes at −38°F. If the temperature drops below this point, the thermometer won't work. Since mercury is poisonous and only can be used to temperatures of −38°F, alcohol thermometers are more popular today. Alcohol doesn't freeze until the temperature drops to −202°F.

As we have seen, different substances boil, melt, or freeze at different temperatures. Atoms and molecules move faster as matter is heated. They move more slowly as matter is cooled. No matter what the substance is, all molecules stop moving at −459°F. We call this point absolute zero. Matter cannot be colder than this temperature.

---

tungsten
A hard metal that has a very high melting point.

mercury
A poisonous silver-colored metal that has a low melting point.

Lesson 19

Practice

Vocabulary in Context ◆ **Choose the word that best completes each sentence.**

1. The metal with the highest melting point is _____ .

2. At 5,000°F, this metal is in the solid _____ .

3. It melts to form a _____ at 6,116°F.

gas
liquid
state
tungsten

Using Context Clues ◆ **Choose the word or words that best complete each sentence. Fill in the circle for your answer.**

4. In the first paragraph on page 99, *vibrate* probably means

   Ⓐ line up in an orderly pattern.
   Ⓑ break down.
   Ⓒ move back and forth.
   Ⓓ melt.

5. In the third paragraph on page 100, a *filament* is

   Ⓐ a metal with a high melting point.
   Ⓑ a metal with a low freezing point.
   Ⓒ a liquid in a light bulb.
   Ⓓ a tiny wire that lights up.

Finding Facts ◆ **Choose the word or words that best complete each sentence. Fill in the circle for your answer.**

6. When H_2O is in its solid state, its molecules are

   Ⓐ different from water.
   Ⓑ different from water vapor.
   Ⓒ lined up in an orderly pattern.
   Ⓓ not moving.

7. The states of matter are

   Ⓐ solid.
   Ⓑ liquid.
   Ⓒ gas.
   Ⓓ all of the above

Check your answers on page 169.

Strategies for SUCCESS

Cause and Effect

When you study chemistry, you need to know the causes and effects of certain actions. Scientists try to explain how one thing happened, then how a second thing happened as a result. The sentence below shows cause and effect.

I took an ice cube out of the freezer, so it melted.

> ❖ **STRATEGY: Look for key words and phrases.**
>
> 1. Some of the words and phrases that show cause and effect are *so, for, because, cause, effect, make, as a result, for this reason, consequently, therefore, when,* and *if*.
>
> 2. Use what you know to figure out what the cause is and what the effect is.

Exercise 1: Read the paragraph and answer the questions.

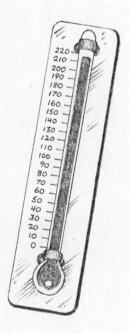

All metals except mercury are solid at room temperature. Mercury is liquid at room temperature. Because of this, it is used in thermometers to show the temperature. However, mercury freezes at −38°F. If the temperature drops below this point, a mercury thermometer won't work. Therefore, thermometers that will be used in very cold areas are made with alcohol.

The phrase "Because of this" tells you that a result follows. "Mercury is liquid at room temperature" is the reason, or the cause, that thermometers are sometimes made with mercury.

1. What causes mercury thermometers to stop working?

   _____

2. What word helped you find the cause? _____

3. Why are thermometers made with alcohol?

   _____

Unit 3

❖ **STRATEGY: Figure out from the context what causes something to happen.**

1. Read and reread the paragraph.
2. Look for clues such as the order in which actions take place.
3. Use what you know to figure out what the cause is and what the effect is.

Exercise 2: Read the paragraphs and answer the questions.

When water is solid ice, all its molecules vibrate slightly and are lined up in an orderly pattern. If you heat the ice, the water molecules move back and forth faster. The warmer the ice becomes, the faster the molecules move. Soon they are moving so fast they can't stay in place any longer. Then the ice melts.

When water, or any solid, melts, it can no longer keep its original shape. Instead, the molecules become rearranged and the liquid takes the shape of the container it is in. As a liquid is poured into a different container, the molecules flow over one another and fall into the new container. The liquid takes the shape of the new container.

1. What causes ice to melt?

_____

_____

2. What change in shape occurs in ice as a result of melting?

_____

3. Why does this change occur?

_____

Check your answers on page 169.

Lesson 20

The Conservation of Matter

When ice melts and turns into water, all the water molecules are still there. There is as much water as there was ice. None of the H_2O molecules has gone anywhere.

When water boils and turns into water vapor, all the water or water molecules still exist. When a pan of water boils on the stove, we can watch the water "disappear." However, when all the water has boiled away and the pot is dry, we cannot see where the water has gone.

Even if we can't see them, the water molecules still exist. Every molecule of H_2O is out there somewhere in the air. This can be proven. Scientists weigh an amount of water and then heat it until it becomes water vapor. They capture all the water vapor. Then they cool and condense the water vapor so it becomes liquid water again. Next they weigh it. It will always weigh the same as it did originally. It will have changed its state from a liquid to a gas and back to a liquid again. However, it will contain the same number of water molecules. Matter will not have been created or lost. This is called the **Law of Conservation of Matter**.

Another example of this law is a log burning in a fireplace. It seems as if the log disappears. It may look as if most of the matter in the log has been destroyed, but it hasn't. It has changed into ash and smoke.

Law of the Conservation of Matter
A rule that says matter cannot be created or destroyed.

Burning is another kind of change that matter can go through.

chemical reaction
A chemical change in a substance.

decomposition
A chemical reaction in which a compound is broken down into two or more simpler substances.

break down
Split into simple substances.

corrosion
The reaction of metals with air.

universe
All existing things, including the earth, the solar system, and the galaxies.

Suppose you could catch all the smoke and water vapor that went up the chimney. You could weigh it, along with the ash that was left after the fire went out. You would find that all the matter in the log is accounted for. The matter in the burning log went through a **chemical reaction**. It was changed into other forms of matter (smoke, water vapor, and ash).

The conservation of matter can be seen in all kinds of chemical reactions. Another kind of reaction is **decomposition**. When a substance decomposes, it **breaks down** to form other substances. The chemical equation below shows a decomposition reaction. (Note: The arrow points away from what reacts and toward what forms.)

$$CaCO_3 \longrightarrow CaO + CO_2$$
calcium carbonate — calcium oxide — carbon dioxide
(limestone) — (lime) — (gas)

When heated, limestone breaks down to form a solid substance called lime. It also produces carbon dioxide gas. Lime, produced in this reaction, can be used to make plaster and cement.

Corrosion is another kind of chemical reaction. You probably have seen rusty cars or street signs. Rust is a common product of corrosion. Rusting occurs when iron reacts with oxygen in air. The substance formed is called iron oxide.

$$2Fe + 3O_2 \longrightarrow Fe_2O_3$$
iron — oxygen gas — iron oxide

There is a certain number of atoms in the **universe**. The number is very large, but that's all the matter that exists. The substances that these atoms are part of may change state. However, the total amount of matter in the universe is always the same.

Lesson 20

Lesson 20

Practice

Vocabulary in Context ◆ **Write the word that best completes each sentence.**

1. The total amount of matter in the _____ remains the same.

2. Burning is an example of a chemical _____ .

> conservation
> reaction
> universe

Using Context Clues ◆ **Choose the words that best complete each sentence. Fill in the circle for your answer.**

3. In the first paragraph on page 105, *ash* probably means

   Ⓐ the floor of a fireplace.
   Ⓑ a kind of chemical reaction.
   Ⓒ the solid matter that is left after something burns.
   Ⓓ a kind of gas given off.

4. In the third paragraph on page 104, *condense* probably means

   Ⓐ to change H_2O to a solid state (ice).
   Ⓑ to change H_2O to a liquid state (water).
   Ⓒ to change H_2O to a gas (water vapor).
   Ⓓ to weigh H_2O in any state.

Finding the Main Idea ◆ **Choose the words that best answer each question. Fill in the circle for your answer.**

5. What is the main idea of the first paragraph on page 105?

   Ⓐ You can weigh ash, water vapor, and smoke.
   Ⓑ When something burns, it disappears.
   Ⓒ Burning changes matter but doesn't destroy it.
   Ⓓ When you burn wood, it changes into smoke, water vapor, and ash.

6. What is the main idea of the last paragraph on page 105?

   Ⓐ There is a certain number of atoms in the universe.
   Ⓑ There is a very large number of atoms in the universe.
   Ⓒ Substances change from one state to another.
   Ⓓ Matter may change state, but it cannot be destroyed.

Check your answers on pages 169–170.

Lesson 21

Fermentation

Nothing smells more appealing than bread fresh from the oven. You may have seen bread dough rising. This is the result of a chemical reaction. The reaction happens in a tiny living thing, **yeast**.

For many centuries, people have used yeast to make bread, wine, and beer. Yeast is a kind of fungus that is found on fruit and feeds on sugar.

You have already learned about burning, decomposition, and corrosion. Another reaction is **fermentation**. Yeast cells use fermentation to live without air and get energy from their food, sugar. The chemical equation for fermentation is shown below.

$$C_6H_{12}O_6 \longrightarrow 2C_2H_5OH + 2CO_2$$
$$\text{sugar} \qquad \text{alcohol} \qquad \text{carbon dioxide gas}$$

Sugar breaks down to form alcohol and carbon dioxide gas. It is this gas that causes bread dough to rise. When bread dough is made, live yeast and sugar are added. The yeast breaks down the sugar. As more carbon dioxide gas is produced, the dough expands. As the dough is baked, the yeast dies and the alcohol evaporates.

yeast
A one-celled fungus that breaks down sugar.

fermentation
A chemical reaction in which sugar is broken down into alcohol and carbon dioxide.

Fermentation is a kind of chemical reaction that causes bread to rise.

Fermentation is also part of wine making. Most wine begins as grape juice or some other fruit juice. Yeast is added to the juice. Then the yeast is given time to ferment the sugar in the juice. Slowly, the sugar is broken down into alcohol and carbon dioxide gas. By this process, the sweet juice is turned into wine. The carbon dioxide is usually allowed to escape. However, if the fermentation is done in a closed container, the gas is trapped. The bubbly result is called sparkling wine or champagne.

Wine-making containers often have special valves. They let carbon dioxide gas escape but keep bacteria from getting in.

hops
The dried, ripe flowers of the hop plant used in brewing beer.

Like wine, beer is the result of fermentation. Yeast is added to a liquid made from mashed grains, water, and dried **hops**. Sugar found in the grain is fermented by the yeast to form alcohol. The carbon dioxide gas given off during fermentation is trapped. This is why beer has bubbles in it.

The alcohol that forms in the reactions described above is called ethanol, or ethyl alcohol. There are many other forms of alcohol that are chemically different from ethanol. These other forms are poisonous and should not be consumed. Labels on rubbing alcohol and other products that contain alcohol warn that the contents are poisonous. Sometimes foul-tasting chemicals are added to alcohol to keep people from drinking it. These **denatured** alcohols also have warning labels on them.

denatured
Made unfit for drinking.

Lesson 21

Practice

Vocabulary in Context ◆ **Write the word that best completes each sentence.**

1. _____ is an organism used to break down sugar in beer making.

2. _____ is a kind of chemical reaction in which ethyl alcohol and carbon dioxide form.

fermentation
hops
yeast

Cause and Effect ◆ **Choose the words that best complete each sentence. Fill in the circle for your answer.**

3. Bread dough rises because

    Ⓐ carbon dioxide gas is produced during fermentation.
    Ⓑ alcohol is produced during fermentation.
    Ⓒ yeast is a fungus.
    Ⓓ the yeast dies during baking.

4. Foul-tasting substances are sometimes added to alcohol

    Ⓐ because carbon dioxide gas is produced.
    Ⓑ to stop fermentation.
    Ⓒ because yeast causes the sugar to break down.
    Ⓓ to make it unfit for drinking.

Finding Facts ◆ **Choose the words that best complete each sentence. Fill in the circle for your answer.**

5. The breakdown of sugar during fermentation produces

    Ⓐ $C_6H_{12}O_6$.
    Ⓑ only alcohol.
    Ⓒ $C_6H_{12}O_6$ and CO_2.
    Ⓓ CO_2 and alcohol.

6. Fermentation produces alcohol

    Ⓐ in bread, wine, and beer making.
    Ⓑ when CO_2 is added to dough.
    Ⓒ after bread dough is baked.
    Ⓓ when grain, water, and hops are mixed.

Check your answers on page 170.

Lesson 22

Hard and Soft Water

hard water
Water that has a high mineral content.

You might think that the only time water is "hard" is when it is frozen. However, there is such a thing as **hard water**. If you've ever lived in an area with hard water, you probably know the problems that it can cause. Hard water makes soap difficult to lather. It also makes soap form a solid scum that forms on sinks and bathtubs.

Calcium deposits in hard water build up and clog pipes.

ion
An atom or group of atoms that have a positive or negative charge.

solution
A mixture that consists of one substance dissolved in another.

When water in the ground flows over rocks, it dissolves some of the rocks' minerals. In water, the minerals break down to form **ions**, which have positive or negative charges.

Calcium ions (Ca^{2+}) are the most common cause of hard water. They come from a kind of rock called limestone. Limestone dissolves in water to form two kinds of ions.

$$CaCO_3 \longrightarrow Ca^{2+} + CO_3^{2-}$$
calcium carbonate   calcium ion   carbonate ion
(limestone)

The water and ions form a **solution**. You can't see the ions in this solution, but they are there.

Unit 3

Sooner or later the solution containing Ca^{2+} enters some homes and water supplies. The ions in the water are not harmful or dangerous to you. When they come in contact with soap, a reaction occurs.

One common kind of soap contains a substance called sodium stearate. It dissolves in water to form a solution of sodium ions and stearate ions. When the stearate ions come in contact with the Ca^{2+} ions in the hard water, a **precipitate** forms. We see that precipitate as soap scum.

calcium ions   +   stearate ions → calcium stearate
(from hard water)   (from soap)   (scum)

The equation above is a word equation. Like a chemical equation, it shows what substances react. It also shows what is produced. This equation uses words instead of symbols to describe the substances.

Hard water is a problem in both homes and industries. When hard water is heated, calcium precipitates may form. These precipitates form crusty layers that clog hot water pipes and boilers. In textile and paper mills, hard water can leave stains on fabric and paper. Problems like this can be avoided if the ions are removed from hard water.

Removing the ions produces **soft water**. Some towns and industries soften hard water by adding chemicals to it. These chemicals react with the ions and cause them to form precipitates. These solids settle out of the water and fall to the bottom of the container. Other chemicals exchange the problem ions for others that don't create problems. This is how home water softeners work.

Soft water makes soap easy to lather. No scum forms on sinks or bathtubs. Some people even say their hair feels softer when they wash it in soft water. However, many people dislike the taste of softened water. For this reason, water softeners are usually connected only to the hot water supply.

---

precipitate
A solid that forms from a solution due to a chemical reaction.

soft water
Water that has a low mineral content.

Lesson 22

Practice

Vocabulary in Context ◆ **Write the word that best completes each sentence.**

1. A hard water ___ _____ contains calcium ions.

2. These ions can react with stearate _____ from dissolved soap to form a _____ .

3. Water that has a low mineral content is called _____ water.

> hard
> ions
> precipitate
> soft
> solution

Finding Facts ◆ **Choose the word or words that best answer each question. Fill in the circle for your answer.**

4. What forms when hard water comes in contact with dissolved soap?

   Ⓐ H_2O
   Ⓑ stearate ions
   Ⓒ calcium stearate
   Ⓓ CO_3^{2-}

5. What is the difference between hard and soft water?

   Ⓐ Soft water contains more minerals.
   Ⓑ Hard water contains more sodium stearate.
   Ⓒ Soft water is liquid and hard water is solid.
   Ⓓ Hard water contains more minerals.

Classifying ◆ **Choose the words that best answer the question. Fill in the circle for your answer.**

6. Which of these problems is related to hard water?

   Ⓐ The water is dangerous to drink.
   Ⓑ Soap scum forms.
   Ⓒ The water tastes unpleasant.
   Ⓓ both A and B

Check your answers on page 170.

Lesson 23

Acid Rain

acid rain
Any rain, snow, or sleet that is acid.

acid
A substance that forms hydrogen ions in water.

pH scale
A measure of how strong an acid or base is.

base
A substance that forms hydroxide ions in water.

Pollution is a fact of life in our world. **Acid rain** is a term you've probably heard in connection with pollution.

The pollution that causes acid rain comes from cars, factories, and power plants. They all burn fuels, which include coal, gasoline, and oil. When fuels are burned, pollutants that contain sulfur are given off. In the air, these pollutants react with gases and with rainwater to form sulfuric acid. This **acid** falls to the ground in acid rain. There are many kinds of acids. Sulfuric acid is strong and harmful. Other acids are not harmful. For example, every time you bite into an orange, you taste citric acid.

Scientists use the **pH scale** to show how strong an acid is. The pH scale runs from 0 to 14. Acids have a pH of less than 7. Notice on the diagram of the pH scale that acids are found to the left. To the right are **bases**. Bases have a pH of more than 7. Substances that are neither acids nor bases, like pure water, have a pH of 7.

The strongest bases are found at the far right on the pH scale. The strongest acids are found at the far left. Acid rain has a pH of less than 5.6. That's almost as strong an acid as vinegar.

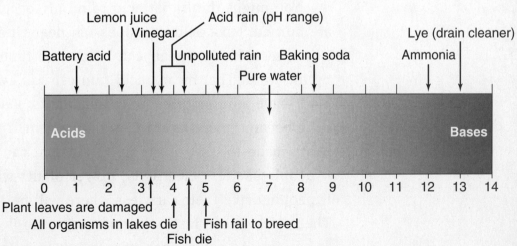

pH Scale

Acid rain can damage stone and metal. It causes chunks of stone to crumble and fall off. As a result, it damages buildings, monuments, and tombstones. It corrodes metal, causing car finishes to pit and rust.

Acid rain damages and kills plants and animals. When acid rain falls on lakes and rivers, it makes them acidic. This has already killed all life in thousands of lakes. Over 15 million acres of forest around the world are dead or dying because of acid rain falling on the trees.

Acid rain caused damage to this statue.

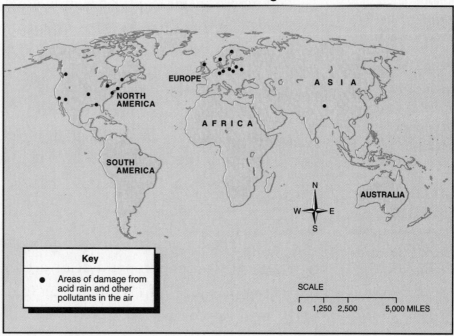

You might think that acid rain only occurs where fuels are burned. Most acid rain does fall near cities and factories. However, acids in the air can be carried thousands of miles by the wind. That means acid rain can fall even in areas that don't have any factories or power plants. Recently people have become more aware of this problem, and efforts are being made to control the pollution that causes acid rain. If people use less electricity, power plants will use less fuel. If people drive their cars less, there will be less pollution. That would mean less acid rain. Acid rain is a problem that is not easy to solve. Each person has to help reduce the amount of pollution in the air.

Lesson 23

Practice

Vocabulary in Context ◆ **Write the word that best completes each sentence.**

1. The damage caused by acid _____ includes corrosion of metal and pitting of stone.

2. A _____ has a pH of greater than 7.

3. An _____ forms H⁺ ions in water.

acid
base
pH
rain

Cause and Effect ◆ **Choose the words that best complete each sentence. Fill in the circle for your answer.**

4. Acid rain forms because

   Ⓐ vinegar is in the air.
   Ⓑ pollutants in the air react with rainwater.
   Ⓒ pure water has a pH of 7.
   Ⓓ the rain has a pH of 5.6.

5. Effects of acid rain include

   Ⓐ citric acid in lakes.
   Ⓑ the burning of fuel.
   Ⓒ the killing of plants and animals.
   Ⓓ the forming of bases in lakes.

Reading a Map ◆ **Use the map on page 114 to choose the answer. Fill in the circle for your answer.**

6. Damage has been caused by acid rain in

   Ⓐ North America and South America.
   Ⓑ Africa and Asia.
   Ⓒ North America and Europe.
   Ⓓ South America and Australia.

7. Based on this map, it is likely that Europe has

   Ⓐ many cities with heavy traffic.
   Ⓑ large amounts of rain.
   Ⓒ many factories.
   Ⓓ both A and C

Check your answers on pages 170–171.

Strategies for SUCCESS

Reading Tables and Graphs

Information can be presented in various ways—sentences, tables, or graphs. To get information from tables and graphs, find out first how they are organized. Then you will have a better idea of what facts are presented and how to find the fact you need to know.

> ❖**STRATEGY: Become familiar with the way the table is organized. Then locate the fact you need.**
>
> 1. Find out how the table is organized. Read the column heads and notice what each row represents.
>
> 2. Find a specific fact in the table by reading across a row and down a column.

Exercise 1: Use the table below to answer the questions.

Some Common Acids		
Name	Chemical Formula	Where It Is Found
Sulfuric acid	H_2SO_4	in acid rain
Acetic acid	CH_3COOH	in vinegar
Hydrochloric acid	HCl	in the stomach
Citric acid	$C_6H_8O_7$	in citrus fruits

1. List the column heads shown in the table.

   _____

   _____

2. What is the chemical formula for citric acid?

   _____

3. Which acid is found in the human body?

   _____

Unit 3

> ❖ **STRATEGY:** Become familiar with the way the graph is organized. Then locate the fact you need.
>
> 1. Find out how the graph is organized by noticing what each side, or axis, represents.
> 2. Notice the scale of each axis.
> 3. Find a specific fact on the graph by reading down one axis and across the other axis.

Exercise 2: Use the graph below to answer the questions.

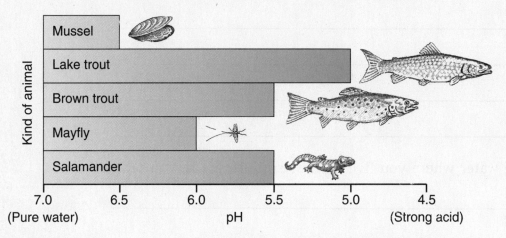

1. What are the two axes on the graph? Which has a scale?

   _____

   _____

2. Which animal shown can survive the strongest acid condition?

   _____

3. Which animal shown is least likely to survive a strong acid condition?

   _____

Check your answers on page 171.

Thinking and Writing

1. Give two examples of how chemical reactions play a part in your life.

   _____
   _____
   _____
   _____

2. Describe a change in state for three different substances. Include each of the three states at least once.

   _____
   _____
   _____
   _____

3. Is the water where you live hard or soft? How can you tell?

   _____
   _____
   _____
   _____

4. What are some things you can do to help cut down on the amount of acid rain that forms?

   _____
   _____
   _____
   _____

Check your answers on page 171.

Unit 4

PHYSICS

Physics is the study of matter and energy and the laws that govern them. If you have ever knocked a glass off the table and watched it crash on the floor, you have experienced physics.

How might you use physics in daily life?

_____

_____

Preview the unit by looking at the titles and pictures. **Write something that you predict you will learn about physics.**

_____

_____

Physics is the study of forces and energy that affect matter. It includes the study of light and sound.

In this unit you will learn about:
- sound
- light
- gravity
- lasers
- nuclear energy

Lesson 24

Sound

sound wave
A pattern of vibrations carried through air, water, and solid objects.

In order to have sound, we need two things. We need someone or something to make the sound. Also, we need someone to hear it. Instead of using the word *sound*, we should use the words *sound waves*. **Sound waves** are real and they can be measured. They don't depend on a hearer. Sound waves are started by vibrations. Take a rubber band and pull it tight. While it is still held tight, pluck the rubber band. You will see the rubber band vibrating, and you will hear sound. That's because the vibrations cause sound waves to start moving.

How do you hear the vibrations from a rubber band? Vibrations push the air in the same way that a stone makes waves when you drop it into a pond. A small stone makes small waves that travel out to the edges of the pond. A large rock makes large waves. In the same way, a small vibration causes small sound waves. A loud noise causes large sound waves. Your vibrating rubber band caused small sound waves to move through the air.

Most of the sounds we hear travel through air. However, sound can also travel through solid objects. Put your ear down on a tabletop and tap on the table. You will hear a sound. Sound travels through water, too. Have you ever been swimming underwater and heard the sound made as someone jumps into the water? As long as there is something for it to travel through, sound will travel.

The table lists some different materials and the speed that sound travels through them. As you can see, sound waves travel faster through solid objects than through air.

Speed of Sound	
Material	Speed in Feet Per Second
Air (68°F)	1,125
Water	4,915
Lucite plastic	8,793
Steel	16,600
Aluminum	16,700
Pyrex glass	18,500

Unit 4

reflection
The effect that happens when light, sound, or heat bounces off a wall or other surface.

decibel
A unit of measurement of volume.

hertz
A unit of measurement of pitch.

If sound travels faster through solid objects, why is it easier for us to hear a person who is talking directly to us than someone who is talking through a wall? This is because sound waves, like rubber balls, bounce off walls and other surfaces and cause a **reflection**. The echoes we hear in a tunnel are the sound waves reflecting off the surface. Not all of the sound waves that hit a wall are reflected. Some are absorbed. Some pass through the wall. What you hear may not be very loud or clear. This is because only a few sound waves actually pass through the wall. Most of them are reflected back to the speaker or absorbed by the wall.

Sound has two major characteristics: volume and pitch. Volume (or loudness) is measured in **decibels** (dB). Pitch (or how high or low a sound is) is measured in **hertz** (Hz).

The louder the sound, the more energy it has. The quietest sound that most people can hear has a volume of 0 decibels. A whisper has a volume of 20 decibels. This is 100 times the energy level of the quietest sound that can be heard. The sound of a moving subway train is about 100 decibels. This sound has 100,000,000,000 times the intensity of the quietest sound that can be heard.

Being exposed to loud sounds can damage your ears. The 100-decibel subway is loud enough to damage the ears of someone who is exposed to this sound for long periods of time. The sound of a jet engine is even louder. This is why baggage handlers and other people who work around jet engines wear ear protection. People who work around other noisy machines also wear ear protection.

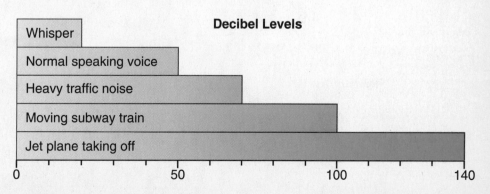

Lesson 24

The diagram below shows four different sounds. Each sound is made up of sound waves. Notice that the waves vary in how tall and how far apart they are. The height of the waves is related to their volume. The higher the sound waves are, the louder the volume is. The distance between the waves is related to their pitch. The closer the waves are, the higher the pitch is.

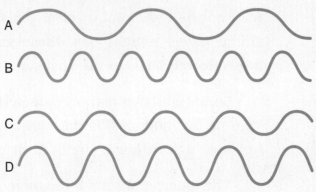

Sounds A and B have the same volume. B has a higher pitch than A, where the waves are farther apart.

Sounds C and D have the same pitch. The distance between the waves is the same. D is louder than C because the height of the wave is greater.

The human ear can hear pitches from as low as 20 Hz to as high as 20,000 Hz. Bats and porpoises can make sounds that are as high as 100,000 Hz. That is five times higher than humans can hear.

The outer ear, the part of the ear we see, receives sound waves and directs them toward the eardrum. Like a real drum, the eardrum vibrates and transmits the sound waves to the inner ear. The inner ear is full of a liquid that is moved by the sound waves. Hairlike structures in the inner ear feel the moving liquid and send messages to the brain. When this happens, we are hearing.

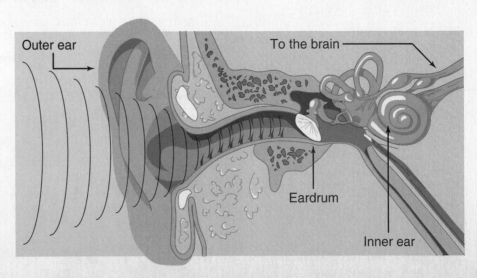

Parts of the ear

Lesson 24

Practice

Vocabulary in Context ◆ Write the word that best completes each sentence.

1. A _____ is a way of measuring sound loudness.

2. _____ occurs when sound waves bounce off a surface.

> decibel
> hertz
> reflection

Cause and Effect ◆ Choose the words that best answer each question. Fill in the circle for your answer.

3. What happens when you pluck a rubber band that is pulled tight?

    Ⓐ You see the rubber band vibrating.
    Ⓑ You hear a sound.
    Ⓒ You see the sound waves.
    Ⓓ both A and B

4. Why are echoes possible?

    Ⓐ Sound waves reflect well off walls and other surfaces.
    Ⓑ Sound travels faster through solid objects.
    Ⓒ Some sound waves are absorbed.
    Ⓓ both B and C

Using Context Clues ◆ Choose the word or words that best complete each sentence. Fill in the circle for your answer.

5. In the third paragraph on page 121, *intensity* probably means

    Ⓐ pitch.
    Ⓑ energy level.
    Ⓒ frequency.
    Ⓓ sound wave.

6. In the last paragraph on page 122, *transmit* probably means

    Ⓐ travel.
    Ⓑ make noise.
    Ⓒ absorb.
    Ⓓ send.

Check your answers on pages 171–172.

Lesson 25

Gravity

gravity
The force that causes all objects in the universe to attract each other. It also helps to keep objects in their places.

mass
The amount of matter an object has.

newton
A unit that measures force.

If you drop a ball, it falls down. If you trip, you fall down, too. The earth's **gravity** is the force that makes these things happen. Gravity holds you on the earth's surface. This force also holds the earth, moon, and planets in their places.

Earth is not the only object that has gravity. This force exists between any two objects. The strength of the force depends on two things: the **mass** of the objects and the distance between them. The more mass the objects have, the greater the force attracting them. Also, the closer together the objects are, the greater the force attracting them.

Your weight is a measure of the force of gravity. When you step on a scale, it measures the force of the pull between you and the earth. The graph below shows how the strength of the pull changes as a person moves farther away from the earth's surface. The greater the distance any object is from the earth, the smaller the force of gravity is on that object. On the graph, force is shown in **newtons** (N). One newton is slightly less than one fourth of a pound.

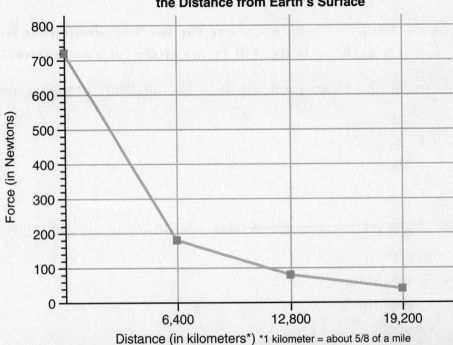

The force of gravity on an astronaut versus the distance from the earth's surface.

Unit 4

If you could travel to the moon, your weight would change. The moon has less mass than the earth. An astronaut on the moon feels only about one sixth the force of gravity that is felt on the earth. An astronaut who weighs 180 pounds on the earth would weigh only 30 pounds on the moon! However, the astronaut would not be thinner or smaller. The astronaut's mass would be the same as it is on the earth. Only the pull of gravity on that mass would be different.

There is a force of attraction between the earth and the moon. Yet the moon does not fall to the earth's surface the way a dropped ball does. The moon is kept circling the earth by a combination of two things: the force of gravity and the moon's tendency to fly off into space.

Imagine tying a rock to a piece of string and whirling the rock around in a circle. If you let go of the string, the rock will fly off in a straight line. You started the rock moving, and you could feel it pull on the string. The string pulled on the rock with **centripetal force**. This force pulled the rock toward the center of its circular path. When you let go of the string, nothing pulled on the rock, so it was free to move away. Gravity is the centripetal force that pulls the moon toward the earth. If gravity didn't exist, the moon would just move out into space and never come back. The earth's gravity acts like a string, keeping the moon from spinning away.

centripetal force
The force that causes things to go toward the center.

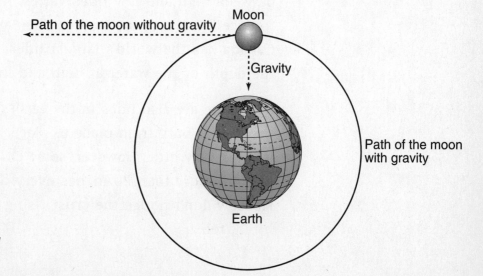

The moon stays in its orbit because the moon's tendency to move out into space is balanced by the pull of the earth's gravity.

Lesson 25

tide
The rise and fall of the level of the sea.

The force of attraction between the earth and the moon has another effect. You have seen that the earth pulls on the moon. The moon pulls on the earth, too. The moon's gravity is not strong enough to pull the earth out of its orbit. However, it does cause the earth to move slightly. The pull of the moon's gravity on the earth causes the **tides**.

As you can see in the diagram below, there is a high tide on the side of the earth facing the moon. On this side of the earth, the water is pulled toward the moon by the moon's gravity. There is also a high tide on the side of the earth facing away from the moon. This high tide is also caused by the moon's gravity. The moon pulls on the earth, pulling it away from the water that is on the side facing away from the moon. This creates a bulge of water on both sides of the earth. It is for this reason that we have two tides each day.

The moon pulls on the oceans and the water rises. This is called high tide.

The parts of the earth that are not having high tides are having low tides. The change in the depth of the water between high and low tides varies. In most places, the difference is only a foot or two. The Bay of Fundy in eastern Canada has the world's largest tides. Here, the difference in the depth of the water at high and low tide is about 45 feet.

There are also tides in the earth's crust. Like ocean tides, these tides vary from place to place. In most places the crust is lifted very little. However, some cities, such as Moscow, rise and fall more than 20 inches every day. People who live in Moscow don't notice the crust rising because they are being lifted, too.

Unit 4

Lesson 25

Practice

Vocabulary in Context ♦ Write the word or words that best complete each sentence.

1. The earth's _____ causes the _____ force that keeps the moon from spinning out into space.

2. Forces are often measured in _____ .

> centripetal
> gravity
> newtons
> rotates

Reading Graphs ♦ Use the graph to answer each question. Fill in the circle for your answer.

3. What would the force of gravity be on an astronaut who is 12,800 km above the earth's surface?

   Ⓐ 12,800 N
   Ⓑ 720 N
   Ⓒ 80 N
   Ⓓ 180 N

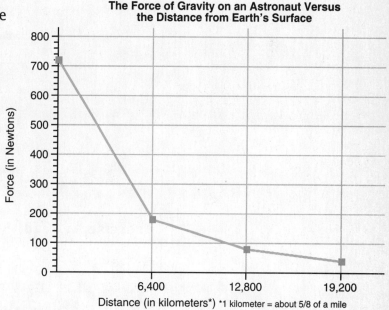

4. How does the force of gravity on the astronaut compare at the earth's surface and at a distance of 19,200 km?

   Ⓐ The force is the same.
   Ⓑ The force is less at the surface.
   Ⓒ The force is about 675 N more at 19,200 km.
   Ⓓ The force is about 675 N more at the surface.

Cause and Effect ♦ Choose the words that best answer the question. Fill in the circle for your answer.

5. What would happen if the earth had no gravity?

   Ⓐ The moon would fall into the earth.
   Ⓑ The moon would move out into space.
   Ⓒ The moon would stay where it is.
   Ⓓ The moon would move faster.

Check your answers on page 172.

Strategies for Success

Drawing Conclusions

Writers don't always say everything directly. If you study the facts and read between the lines, you can get a lot of information from a passage. You can use all this information to draw your own conclusions. A conclusion is a judgment you can make after studying all the facts you have. Read the following example.

> A small vibration causes small sound waves.
> A loud noise causes large sound waves.

From these sentences, you can conclude that the size of a sound wave depends on the strength of the vibration that caused it.

> ❖ **STRATEGY: Study the facts. Then think it through.**
>
> 1. As you read a passage, ask yourself: What are the facts?
>
> 2. Then look for ideas you find out by reading between the lines.
>
> 3. Ask yourself: What do the facts and ideas in the passage tell me?

Exercise 1: Read the paragraph. Underline the facts.

How do we hear? The outer ear receives sound waves and directs them toward the eardrum. Like a real drum, the eardrum vibrates and passes the sound waves to the inner ear. The inner ear is full of a liquid that is moved by the sound waves. Hairlike structures in the inner ear feel the moving liquid and send messages to the brain. These messages are what we call hearing.

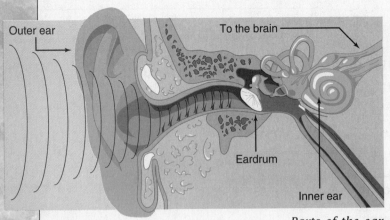

Parts of the ear

Unit 4

Exercise 2: Read the paragraph on page 128 again. What do all the facts about the ear have in common? Fill in the circle for your answer.

  Ⓐ They are all about how the brain works.
  Ⓑ They are all about how hearing occurs.
  Ⓒ They are all about liquids.
  Ⓓ both A and C

Exercise 3: From the paragraph and diagram on page 128, what conclusion can you draw about the purpose of the ear? Fill in the circle for your answer.

The purpose of the ear is to
  Ⓐ make sounds so you can hear.
  Ⓑ collect sounds so you can hear.
  Ⓒ block loud sounds so you can't hear them.
  Ⓓ both A and C

Exercise 4: Read the paragraph below and answer the questions.

> We all live with gravity. It's part of our lives. Because of gravity, when we trip, we fall down and not up. Because of gravity, people, buildings, trees, and the oceans do not fly off into space when the earth turns.

1. What do you think would happen if there were no gravity on the earth?
    Ⓐ Dropped objects would fall slowly.
    Ⓑ Objects would fly off into space.
    Ⓒ Thrown objects would fall down more quickly.
    Ⓓ both A and B

2. What information from the paragraph helped you draw this conclusion?

    A.

    B. _____

Check your answers on page 172.

Lesson 26

Light

Light and sound have some things in common. Like sound, light travels in waves. Unlike sound, light waves can travel through a vacuum or outer space—places with no air. If light waves didn't travel through space, we wouldn't be able to receive any light from the sun.

Light has two basic characteristics: brightness and color. A bright light, like a loud sound, will have a high wave. A dim light will have a smaller wave.

Light of different colors has different **wavelengths**. The wavelength tells if a wave is long or short. This is similar to the way pitch is measured in sound waves. Blue and violet light have the shortest wavelengths of all the colors. Red light has the longest wavelength. The wavelengths of yellow and orange light are in the middle. White light such as sunlight contains all the different wavelengths. In other words, white light contains all the colors of light. These colors form the **spectrum**.

When white light passes through a **prism**, the prism spreads out the colors, and we can see each one separately. The prism bends the light waves. Some of the light waves bend more than others. Violet light waves bend the most. Red light waves bend the least.

wavelength
The distance between the top of one wave and the top of the next.

spectrum
The different colors that white light breaks down into: red, orange, yellow, green, blue, indigo, and violet.

prism
A piece of clear glass that bends white light, showing the spectrum.

When white light passes through a prism, the prism spreads out the colors, and we can see each one separately.

Unit 4

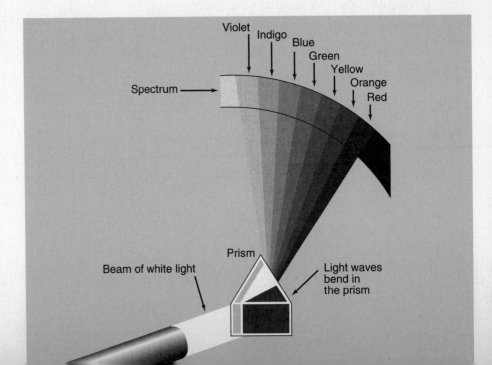

Sometimes raindrops act like tiny prisms. The raindrops break up sunlight into spectrums called rainbows. The diagram below shows that rainbows are formed when the sun is behind you and the rain is in front of you.

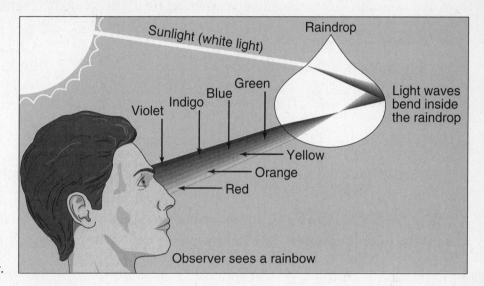

You see a rainbow when raindrops bend sunlight and separate it into different colors.

Have you ever noticed that, during a thunderstorm, you see the lightning a few seconds before you hear the thunder? Lightning and thunder happen at the same time. However, the light reaches you much faster than the sound waves do. The speed of light is 186,000 miles per second. The speed of sound in air is 1,128 feet per second. That's only about one fifth of a mile per second.

You can use this difference to estimate how far away a storm is. When you see lightning, count the seconds until you hear the thunder. It takes five seconds for sound to travel one mile, so if you count five seconds, you know that the lightning was one mile away.

Even though light travels at 186,000 miles a second, it takes many years for the light of a star to reach the earth. For this reason, scientists measure distances in outer space in **light years**. A light year is almost 6 **trillion** miles. The closest star to the earth is four and one-half light years away. Some stars are 80,000 light years away. That means that the light from the stars we see at night left those stars 80,000 years ago.

light year
The distance that light travels in one year.

trillion
1,000,000,000,000.

Lesson 26

Lesson 26

Practice

Vocabulary ◆ **Write the word that best completes each sentence.**

1. A _____ separates white light into different colors.

2. A light year is a distance of 6 _____ miles.

3. The different colors that make up white light are called the _____ .

4. Red light has a longer _____ than violet light.

pitch
prism
spectrum
trillion
wavelength

Drawing Conclusions ◆ **Choose the word or words that best complete each sentence. Fill in the circle for your answer.**

5. Light and sound

  Ⓐ are completely different.
  Ⓑ are alike in some ways.
  Ⓒ travel at the same speed.
  Ⓓ are exactly alike.

6. In order to see a rainbow, we need

  Ⓐ a glass prism.
  Ⓑ white light.
  Ⓒ raindrops.
  Ⓓ both B and C

Cause and Effect ◆ **Choose the words that best complete the sentence. Fill in the circle for your answer.**

7. When white light passes through a prism,

  Ⓐ some of the light waves bend more than others.
  Ⓑ the light waves disappear.
  Ⓒ we can see each color of the spectrum separately.
  Ⓓ both A and C

Unit 4

Check your answers on pages 172–173.

Lesson 27

Lasers

laser
A device that gives off light waves that are all exactly alike.

Some people may think that **lasers** sound like science fiction, but in fact they have become a part of our everyday lives. At the supermarket, lasers read the price codes on your groceries. Your library may use lasers to check out books. Your eye doctor or dentist may have used a laser to operate on you or someone you know. A laser is a device that gives off a special kind of light.

How is laser light different from other kinds of light? To answer this, let's compare the light waves in the table below. Remember that wavelength is the distance from the top of one wave to the top of the next wave. White light is made up of waves with many different wavelengths. Light of one color, such as red or blue, has waves that are similar but not exactly the same. Laser light is made up of waves that all have the same wavelength.

Laser light also differs in the way the waves line up. In other lights, the tops and bottoms of the waves are not lined up. With laser light, all the waves are lined up. Laser light is said to be *in phase*.

Light	Wave Forms	Description
White light		Many different wavelengths, out of phase
Red light		Slightly different wavelengths, out of phase
Red laser light		Same wavelengths, in phase

Lesson 27

133

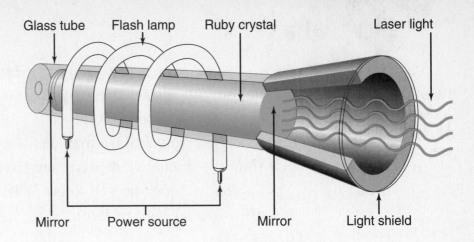

A ruby laser

Optical fibers

optical fiber
Thin fibers of glass or plastic used to send information.

hologram
A three-dimensional image made using laser light.

The diagram shows a simple laser. Light from a flashing lamp is used to give extra energy to the atoms in a ruby. When atoms are given extra energy, they are said to become excited. Then they give off light. This light bounces back and forth between two mirrors and excites more atoms. More and more light is given off. The light becomes very intense and shoots off of one mirror as a laser beam. In addition to rubies, other solids, liquids, and gases can be excited to give off laser light.

Lasers have many uses. They are used to send telephone conversations and other information on **optical fiber** cables. In surveying land, they are used as range finders to measure distances. In the field of medicine, lasers can cut tissue and join it back together. They are especially helpful in brain and eye surgery. Lasers can be used to treat skin cancer, remove tattoos, cut hard metals, and drill diamonds. They make colorful light displays for music concerts and shows. Also, lasers are used to make the **holograms** found on credit cards and decals.

Have you ever wondered how lasers at checkout lines work? Each product in the store is marked with a price code made of black and white bands. The store's computer knows what price each code stands for. When you check out, the clerk passes the price code across a laser. The laser light bounces off the code. A computer senses the light, reads the code, and tells you on a screen what each item costs. Then it adds up your purchases.

Unit 4

Lesson 27

Practice

Vocabulary in Context ◆ **Write the word that best completes each sentence.**

1. The device that gives off light in phase is called a _____ .

2. Information travels as laser light on _____ fiber cables.

3. A _____ is a three-dimensional image created by using light in phase.

> hologram
> laser
> optical
> wavelength

Finding Facts ◆ **Choose the words that best complete each sentence. Fill in the circle for your answer.**

4. Laser light is
   - Ⓐ waves of different wavelengths.
   - Ⓑ not in phase.
   - Ⓒ white light.
   - Ⓓ waves that are of the same wavelength and in phase.

5. Laser light is used
   - Ⓐ to find distances in surveying.
   - Ⓑ for reading lamps.
   - Ⓒ in eye surgery.
   - Ⓓ both A and C

Cause and Effect ◆ **Choose the word or words that best complete the sentence. Fill in the circle for your answer.**

6. When a ruby's atoms are given extra energy, they
   - Ⓐ burn.
   - Ⓑ become excited.
   - Ⓒ give off light.
   - Ⓓ both B and C

Check your answers on page 173.

Strategies for SUCCESS

Predicting Outcomes

An outcome is like an effect. It is something that will happen as a result of something else that happens. When you predict, you make an informed guess about what is likely to occur. You base your prediction on what you already know.

> ❖**STRATEGY:** Review what you already know. Then apply it to the new situation.
>
> 1. Think about what has happened before in similar situations.
> 2. Think of facts that you can apply to this situation.
> 3. Ask yourself: What is the best guess I can make based on what I know?

Exercise 1: Read the text below and study the diagram to find out how rainbows form.

Sometimes raindrops act like tiny prisms. The raindrops break up sunlight into spectra called rainbows. The diagram below shows that rainbows are formed when the sun is behind you and the rain is in front of you.

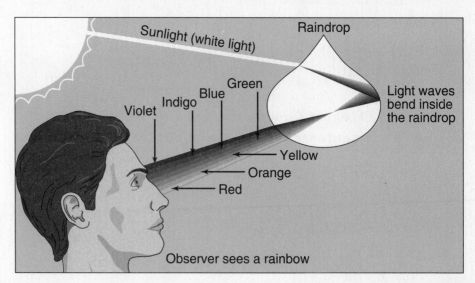

You see a rainbow when raindrops bend sunlight and separate it into different colors.

Unit 4
136

The text and diagram give the following facts.
- Sunlight can be broken up by raindrops into rainbows.
- You see a rainbow when the sun is behind you and the rain is in the sky in front of you.

Predict whether you are likely to see a rainbow in each of the following situations. Explain your predictions. Write your answers below.

1. You are watching the sun rise above a faraway hill. Watching the hill, you realize it has started raining where you are.

_____

_____

2. The sky has been completely covered with clouds all day. You set out for a walk anyway and get caught in the rain.

_____

_____

Exercise 2: **Read the paragraph below and answer the question.**

Light and sound have some things in common. Like sound, light travels in waves. Unlike sound, light waves can travel through a vacuum or outer space—places with no air. If light waves didn't travel through space, we wouldn't be able to receive any light from the sun.

Your friend predicts that, if a nearby star exploded, we would see light from the explosion, but we wouldn't hear it. Do you agree with this prediction? Explain your answer.

_____

_____

Check your answers on page 173.

Lesson 28

Nuclear Energy

Think of all the devices you use that run on electricity. Radios and TVs use electricity. So do lamps, microwave ovens, hair dryers, and clocks. If you're like most people, you probably don't stop to think about where this important form of **energy** comes from. Electricity comes from a power plant, which uses one kind of energy to produce electrical energy.

Your power company may get energy to make electricity from moving water in a waterfall. The company may get energy by burning a fuel, such as coal or oil. Some power companies use nuclear energy, which is the energy given off when atoms are split.

Most of the nuclear energy that is used to make electricity comes from uranium. This element is made up of the heaviest atoms found in nature. Each uranium atom consists of 92 protons in the nucleus and 92 electrons spinning around it. The nucleus is made up of protons and neutrons. Scientists discovered that when one more neutron is added to the nucleus, the nucleus splits and energy is released and **fission** occurs.

Fission doesn't take place easily in nature, but once it starts, it produces extra neutrons. These neutrons are absorbed by nearby uranium atoms. Then these atoms split, causing a **chain reaction**.

Look at the drawing of the dominoes. The dominoes are standing in a row. Behind each domino is another one. When the first one is pushed over, all the rest fall down. This is an example of a chain reaction.

An atomic chain reaction is more complicated. It's more like the chain reaction shown in the drawing on page 139. Behind each domino are two dominoes. When the first is pushed over, a very quick chain reaction takes place. Even though there are more dominoes in this chain reaction, it

---

energy
The ability to cause changes in matter.

fission
The splitting of a nucleus.

chain reaction
A series of events in which each event affects what happens next.

When the first domino is pushed over, all the rest fall down.

Unit 4
138

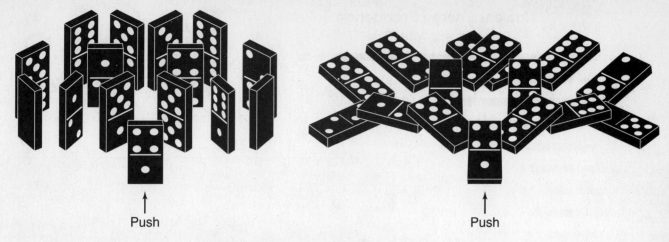

When there are two dominoes behind each domino, a very quick chain reaction takes place.

takes less time for all of them to fall. If this were the splitting of a uranium atom, the first fission would produce two nuclei, the second fission would produce four, and so on.

A small lump of pure uranium contains trillions of atoms. When all of these atoms are forced to split in a fraction of a second, the resulting explosion is enormous. Entire cities and the people living in them can be destroyed by such an atomic bomb.

An atomic bomb is a runaway chain reaction. In order for atomic chain reactions to be useful rather than destructive, they have to be controlled. A **nuclear reactor** is designed to control chain reactions. Here, rods of uranium are placed in special materials. These materials, called moderators, slow down the chain reaction. They do so by capturing some of the neutrons produced when the nuclei split. Materials such as water and graphite are good moderators.

Between the uranium rods there are control rods. Control rods are moved in and out of the moderator. When moved in the moderator, these rods absorb extra neutrons and help control the chain reaction. When the control rods are moved out, more and more energy is given off by fission.

nuclear reactor
Where controlled atomic chain reactions take place.

Lesson 28

Nuclear Energy Production
Percentage of World Total

- United States: 27.4
- France: 15.1
- Japan: 11.4
- Germany: 8.6
- Canada: 4.6
- Sweden: 4.1
- United Kingdom: 3.3
- Belgium: 2.5
- Spain: 2.4
- South Korea: 2.4
- Czechoslovakia: 1.3
- Switzerland: 1.3
- Finland: 1.2
- Bulgaria: 0.7
- Hungary: 0.7
- Argentina: 0.4
- Others: 1.1

The U.S.S. Ohio *nuclear submarine*

Nuclear reactors give off large amounts of energy in the form of heat. This heat is used to boil water and make steam. The steam runs turbines that generate electricity. In this way, nuclear power plants supply energy to homes and industries all over the world. By far the largest producer of nuclear energy in the world is the United States.

Smaller nuclear reactors supply energy to submarines, surface ships, and spacecrafts such as *Voyager*. Submarines powered by nuclear generators can stay underwater for long periods of time. They have been used to do research under the ice of the Arctic Ocean.

Unit 4

Lesson 28

Practice

Vocabulary in Context ◆ **Write the word that best completes each sentence.**

1. When fission occurs very quickly, a runaway _____ reaction occurs.

2. _____ takes place when an extra neutron is added to a uranium nucleus.

chain
fission
reactor

Cause and Effect ◆ **Choose the words that best complete each sentence. Fill in the circle for your answer.**

3. When one more neutron is added to a uranium atom,

   Ⓐ the nucleus splits.
   Ⓑ energy is released.
   Ⓒ cities are destroyed.
   Ⓓ both A and B

4. When trillions of uranium atoms are split in a fraction of a second,

   Ⓐ a huge explosion takes place.
   Ⓑ the uranium stays the same.
   Ⓒ cities are destroyed.
   Ⓓ all of the above

Predicting Outcomes ◆ **Choose the words that best answer the question. Fill in the circle for your answer.**

5. What is likely to happen in a nuclear reactor if the moderator failed and the control rods were removed?

   Ⓐ A controlled chain reaction would occur.
   Ⓑ Uranium atoms would split very slowly.
   Ⓒ A runaway chain reaction would occur.
   Ⓓ Trillions of atoms would form.

Check your answers on page 173.

Lesson 29

Physics Explains Why

Many of the things that happen every day can be explained by understanding physics. For example, why is the sky blue? Why are clouds white? The answers have to do with the properties of light and air.

You have learned that light travels in waves. It also travels in a straight line until it hits something new. Sometimes light passes into and through the new material it hits. This is what happens when light hits prisms and raindrops. Other times light bounces off what it hits and gets scattered in a new direction.

When sunlight enters the earth's atmosphere, it hits molecules in the air. Remember that white light such as sunlight contains all the different colors of light. The molecules in air are just the right size to scatter the shorter wavelengths of blue and violet light. When sunlight hits the air, blue and violet waves are scattered all over the sky and downward. No matter where in the sky you look, blue and violet waves are coming toward you and so you see a blue sky.

Why don't clouds look blue, too? The molecules of ice and water found in clouds are much bigger than the molecules in air. These large molecules can scatter light of all wavelengths. When all the wavelengths together are scattered down to the earth by clouds, you see white light.

Physics helps to explain why you see blue skies and white clouds.

Unit 4

Physics can also help explain thunder. We've discussed why you see lightning before you hear thunder. Why do you think you hear claps and rumbles of thunder? The answer has to do with the properties of light and sound.

You may have noticed that standing in the sun makes you feel warmer. This is because light can heat things. When a gas is heated by light, it spreads out, or **expands**. When the gas cools off, it **contracts**.

When lightning flashes, it heats the air around it. This air expands quickly. After that, the air cools and contracts quickly. These changes make the air molecules vibrate back and forth, and sound waves form. You hear those sound waves as a clap of thunder.

Not all the thunder from a single flash of lightning reaches you at the same time. The part of the lightning flash nearest you produces sound waves that reach you first. You hear this as the first thunderclap. Sound waves from parts of the lightning flash farther from you and reach you later. You hear this as rumbles of thunder.

expand
Spread out.

contract
Move together.

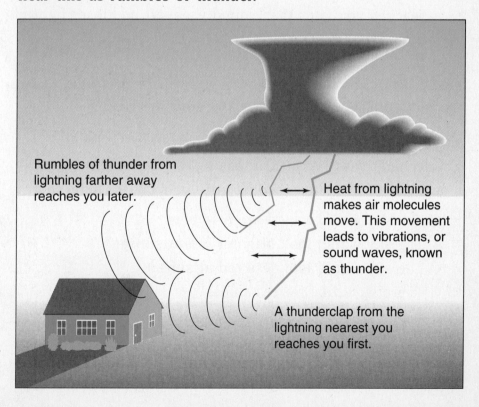

The diagram shows why you hear claps and rumbles of thunder.

Lesson 29

Lesson 29

Practice

Vocabulary in Context ◆ **Write the word that best completes each sentence.**

1. Lightning heats air and causes it to _____ .

2. A gas _____ when it cools down.

> contracts
> expand
> thunder

Cause and Effect ◆ **Choose the words that best complete each sentence. Fill in the circle for your answer.**

3. The sky appears blue because

   Ⓐ air molecules expand when heated.
   Ⓑ air molecules scatter the shorter wavelengths of light.
   Ⓒ water molecules scatter all wavelengths of light.
   Ⓓ both B and C

4. You hear rumbles of thunder after a thunder clap because

   Ⓐ air molecules contract when cooled.
   Ⓑ air molecules expand when heated.
   Ⓒ sound waves from lightning that's farther away from you reaches you later than sound waves from closer lightning.
   Ⓓ both A and B

Predicting Outcomes ◆ **Choose the words that best answer the question. Fill in the circle for your answer.**

5. Dust particles are the right size to scatter the longer wavelengths of red and orange light. What might happen as a result of increased dust in the air due to pollution?

   Ⓐ The sky would look a brighter blue.
   Ⓑ The sky would look brighter red at sunset.
   Ⓒ The sky would look less red at sunset.
   Ⓓ There would be fewer clouds.

Check your answers on pages 173–174.

Thinking and Writing

1. In what other situations (other than rainbows) have you seen a spectrum of colors?

   _____
   _____
   _____

2. Do you think nuclear fission has produced both positive and negative results for people? Explain your answer.

   _____
   _____
   _____

3. Describe something you experienced that could be called a chain reaction. Explain your answer.

   _____
   _____
   _____

4. You have seen that loud sounds can damage a person's hearing. Do you think there should be laws to control loud sounds, such as loud radios? Explain your answer.

   _____
   _____
   _____

Check your answers on page 174.

Check What You've Learned

Check What You've Learned will give you an idea of how well you've learned to understand science content using the skills in this book.

You will read passages, graphs, and maps followed by one or more multiple-choice questions. There is a total of 20 questions. There is no time limit.

Read each passage and question carefully. Fill in the circle for the best answer.

Questions 1–2 are based on the following paragraph.

> The water in streams and rivers carries sand, rocks, mud, and other suspended material. The faster the water moves, the more suspended material it can carry. When the water slows down, some of this material drops out. It is deposited as sediment at the bottom or on the shores of the river or stream.

1. Suspended material drops out of the water of a river when

    Ⓐ the water in the river moves faster.
    Ⓑ the water in the river moves more slowly.
    Ⓒ there is sediment on the shores of the river.
    Ⓓ the suspended material rises to the surface.

2. Sediment is

    Ⓐ sand, rocks, and mud that have been deposited at the bottom of a river.
    Ⓑ sand, rocks, and mud that are suspended in the water of a river.
    Ⓒ a river that moves very slowly.
    Ⓓ the shores of a river that moves very fast.

Questions 3–5 are based on the following paragraph.

Every year trees produce a double layer of wood. They produce a light-colored layer in the spring. Then later, in the summer, they produce a darker-colored layer. If you cut down a tree and look at the stump, you will see the rings that the layers form. Each pair of light and dark rings shows one year that the tree has lived. The older rings are on the inside, and the younger rings are on the outside. Rings vary in thickness. For example, rings that grow in years when there is a lot of rain are thicker than rings that grow in dry years.

3. What is the main idea of this paragraph?

   Ⓐ The oldest rings are on the inside of a tree.
   Ⓑ The rings on a tree vary in thickness.
   Ⓒ Every year trees produce a double layer of wood.
   Ⓓ Trees produce dark-colored rings in the summer.

4. You can conclude that

   Ⓐ trees grow more during wet years.
   Ⓑ trees grow more in dry years.
   Ⓒ trees don't produce rings in dry years.
   Ⓓ trees produce more rings in wet years than in dry years.

5. If a tree has 20 pairs of light-colored and dark-colored rings, you know that

   Ⓐ it's 40 years old.
   Ⓑ it's 20 years old.
   Ⓒ it grew very fast.
   Ⓓ the weather was dry.

Questions 6–8 are based on the following map.

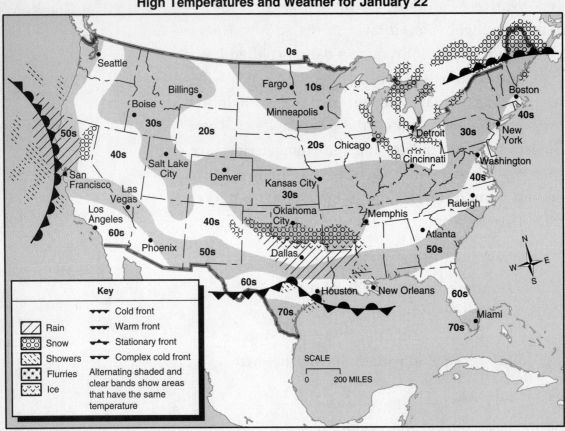

6. In which of the following cities is it raining?

   Ⓐ Minneapolis
   Ⓑ Denver
   Ⓒ Atlanta
   Ⓓ Dallas

7. What is the high temperature in Chicago?

   Ⓐ in the 10s
   Ⓑ in the 20s
   Ⓒ in the 30s
   Ⓓ in the 40s

8. What is the weather like in San Francisco?

   Ⓐ rainy, in the 50s
   Ⓑ rainy, in the 40s
   Ⓒ snowing, in the 30s
   Ⓓ clear, in the 20s

Questions 9–11 are based on the following passage.

When two surfaces rub together, there is friction. Friction slows things down and makes work harder. When the parts of a machine rub together, they cause a lot of friction. Friction makes machines wear out and it wastes some energy. The more friction there is, the more energy the machine needs to run. If the machine is a washing machine or a fan, friction makes it use more electricity. That makes the machine cost more to run.

You can reduce friction in a machine by oiling the gears and other moving parts. The oil makes the parts slippery so they don't rub together. When friction is reduced, less energy is wasted.

9. If there is a lot of friction in a machine, it

    Ⓐ needs more energy to run than a machine that is well oiled.
    Ⓑ must have a motor.
    Ⓒ is more inexpensive to run than a machine that has less friction.
    Ⓓ has no parts that rub together.

10. You can conclude that keeping machines oiled properly

    Ⓐ wastes energy.
    Ⓑ makes them last longer.
    Ⓒ causes more friction.
    Ⓓ makes them cost more to run than machines that are not oiled.

11. The main idea of this passage is that

    Ⓐ friction makes things rub together.
    Ⓑ friction wastes energy.
    Ⓒ friction makes machines break.
    Ⓓ friction can be reduced.

Questions 12-14 refer to the following paragraph and graph.

The volume of a gas decreases when the pressure on it increases, if the temperature remains the same. The volume of a gas increases when the pressure on it decreases, if the temperature remains the same. This relationship is shown in the graph.

Relationship of Volume and Pressure of a Gas When Temperature Remains the Same

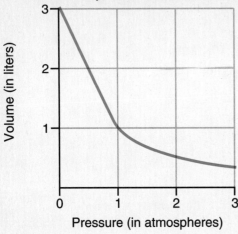

12. What is the unit of measure for pressure on this graph?

    Ⓐ volume

    Ⓑ gas

    Ⓒ liter

    Ⓓ atmosphere

13. When a gas has a volume of 1 liter, what is the pressure according to the graph?

    Ⓐ 0 atmosphere

    Ⓑ 1 atmosphere

    Ⓒ 2 atmospheres

    Ⓓ 3 atmospheres

14. What happens to a gas as the pressure increases?

    Ⓐ The gas takes up more space.

    Ⓑ The gas takes up less space.

    Ⓒ The gas takes up the same amount of space.

    Ⓓ The gas has a volume of 1 liter.

Questions 15–17 refer to the following paragraph and diagram.

The amount of daylight affects the way plants grow. Short-day plants begin to flower when the days are short, as in the autumn. Long-day plants begin to flower when the days are long, as in the summer. When you try to grow long-day plants during a short-day season, they do not grow well. Similarly, when you grow short-day plants during a long-day season, they do not grow well.

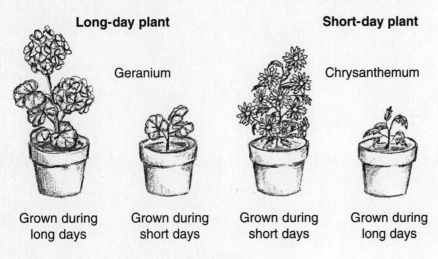

15. When does a geranium bloom?

    Ⓐ during the short days of winter
    Ⓑ during the long days of summer
    Ⓒ during the short days of autumn
    Ⓓ at any time of year

16. At what time of year would you expect to see chrysanthemum flowers?

    Ⓐ June
    Ⓑ July
    Ⓒ August
    Ⓓ September

17. How would you classify a very short geranium that does not flower?

    Ⓐ long-day plant grown during long days
    Ⓑ long-day plant grown during short days
    Ⓒ short-day plant grown during short days
    Ⓓ short-day plant grown during long days

Questions 18–19 are based on the following paragraph.

Metals have many of the same properties, or characteristics. First, they all have luster—they are shiny. Second, all of them, except mercury, are solid at room temperature. Third, metals are conductors of electricity. They allow electricity to pass through them. Metals are also good conductors of heat, so most pots and pans used for cooking are made of aluminum or copper.

18. In this paragraph, *luster* means

    Ⓐ hard.
    Ⓑ shiny.
    Ⓒ used for cooking.
    Ⓓ a conductor.

19. What would you expect aluminum and copper to be used for?

    Ⓐ insulation
    Ⓑ safety shoes
    Ⓒ electrical wires
    Ⓓ potholders

Question 20 is based on the following paragraph.

The animal kingdom is divided into 30 major groups based on body features. For example, animals with backbones, called vertebrates, are in one group. All the other major groups are made up of animals without backbones, called invertebrates.

20. Which of the following animals is a vertebrate?

    Ⓐ cat
    Ⓑ worm
    Ⓒ octopus
    Ⓓ housefly

When you finish *Check What You've Learned*, check your answers on pages 174–175. Then complete the chart on page 153.

Check What You've Learned

The chart shows you which skills you should go back and review. Reread each question you missed. Then look at the appropriate pages of the book for help in figuring out the right answers.

Skills Review Chart

Skills	Questions	Pages
The test, like this book, focuses on the skills below.	Check (√) the questions you missed.	Review what you've learned in this book.
Reading a Diagram	____ 15 ____ 16	UNIT 1 ◆ Pages 15–58 Strategy for Success Pages 28–29
Finding the Main Idea	____ 3 ____ 11 ____ 17	UNIT 1 ◆ Pages 15–58 Strategy for Success Pages 40–41
Classifying	____ 2 ____ 20	UNIT 1 ◆ Pages 15–58 Strategy for Success Pages 56–57
Reading a Map	____ 6 ____ 7 ____ 8	UNIT 2 ◆ Pages 59–92 Strategy for Success Pages 72–73
Using Context Clues	____ 1 ____ 18	UNIT 2 ◆ Pages 59–92 Strategy for Success Pages 90–91
Cause and Effect	____ 9 ____ 12	UNIT 3 ◆ Pages 93–118 Strategy for Success Pages 102–103
Reading Tables and Graphs	____ 13 ____ 14	UNIT 3 ◆ Pages 93–118 Strategy for Success Pages 116–117
Drawing Conclusions	____ 4 ____ 5	UNIT 4 ◆ Pages 119–145 Strategy for Success Pages 128–129
Predicting Outcomes	____ 10 ____ 19	UNIT 4 ◆ Pages 119–145 Strategy for Success Pages 136–137

Glossary

260°F, −280°F A temperature of 260 degrees Fahrenheit, a temperature of minus 280 degrees Fahrenheit. *page 60*

acid A substance that forms hydrogen ions in water. *page 113*

acid rain Any rain, snow, or sleet that is acid. *page 113*

algae The simplest green plants without roots, stems, or leaves. *page 50*

antibiotic A drug that stops the growth and reproduction of bacteria. *page 37*

antibody A chemical made by the immune system to destroy a particular germ. *page 37*

artery A blood vessel that carries blood away from the heart. *page 20*

atmosphere All the air surrounding the earth. *page 60*

atom The smallest particles of an element. *page 95*

atomic number The number of protons in an atom. *page 95*

atomic structure The number of protons, neutrons, and electrons that make up an atom. *page 96*

bacteria Tiny one-celled organisms. They live in water, soil, air, plants, animals, and people. *page 36*

base A substance that forms hydroxide ions in water. *page 113*

blood vessel A tube that blood flows through. *page 20*

blowtorch A tool that uses a mixture of gas and air under pressure. It makes a very hot flame. *page 78*

bovine growth hormone (bGH) A substance that makes cows produce milk. *page 54*

brain stem The part of the brain that controls automatic life processes. *page 24*

brain The part of the body that controls all its activities. *page 24*

break down Split into simple substances. *page 105*

capillary A very small blood vessel. *page 20*

carbon dioxide A colorless, odorless gas produced when fuel is burned. *page 62*

carnivore An animal that eats only meat. *page 47*

cell The basic part of any living thing. All plants and animals are composed of one or more cells. *page 30*

centripetal force The force that causes things to go toward the center. *page 125*

cerebellum The part of the brain that coordinates muscle activities. *page 24*

cerebrum The part of the brain in which thinking occurs. *page 24*

chain reaction A series of events in which each event affects what happens next. *page 138*

chemical reaction A chemical change in a substance. *page 105*

chemistry The study of matter. *page 94*

cholesterol A type of fat found in the human body and in animal foods. *page 22*

circulatory system The heart and all the blood vessels. *page 20*

compost A mixture of organic matter used to enrich soil. *page 88*

compound A substance made of two or more elements. *page 96*

conclusion The decision on whether the hypothesis is true or false. *page 18*

conservation Keeping something safe. *page 43*

contract Move together. *page 143*

core The center. *page 70*

coronary artery disease A disease that causes the arteries to become blocked. *page 22*

corrosion The reaction of metals with air. *page 105*

cortex The part of the cerebrum that thinks and stores information. *page 24*

counter-clockwise Moving in a circle from right to left. *page 83*

crater A funnel- or bowl-shaped pit at the top of a volcano. *page 69*

crust A hard outer covering. *page 69*

cubic mile An area one mile long, one mile wide, and one mile high. *page 61*

current A movement of water in the ocean. It's like a river. *page 65*

DDT A chemical that kills insects. *page 54*

decibel A unit of measurement of volume. *page 121*

decomposition A chemical reaction in which a compound is broken down into two or more simpler substances. *page 105*

denatured Made unfit for drinking. *page 108*

ecosystem A community where plants, animals, and climate are related and in balance. *page 42*

electron A negatively charged particle in an atom. *page 95*

element A substance that can't be broken down into simpler substances. *page 94*

endangered species Any type of animal in danger of extinction. *page 48*

energy The ability to cause changes in matter. *page 138*

environment Everything that surrounds us, including air, water, soil, plants, and animals. *page 53*

erupt Explode violently. *page 61*

eruption An explosion of ash, steam, and lava from a volcano. *page 68*

expand Spread out. *page 143*

experiment A method used to test a hypothesis. *page 17*

extinct No longer in existence. *page 46*

eye The calm area in the middle of a storm. *page 83*

fault A crack in rock along which movement occurs. *page 76*

fermentation A chemical reaction in which sugar is broken down into alcohol and carbon dioxide. *page 107*

fern Green plants with roots, stems, and leaves but no seeds. *page 50*

fiber Thin threads that make up fabric and other materials. *page 88*

fiberfill A material made from fibers that is used to create bulk or warmth. *page 88*

fission The splitting of a nucleus. *page 138*

food chain A cycle in which plants are eaten by animals, who are eaten by other animals. *page 54*

fossil The remains of a once-living thing. *page 46*

front Boundary between types of weather. *page 73*

fruit The part of a plant that contains seeds. *page 50*

fumarole A hole in the ground where smoke and gases escape. *page 79*

gas The state of matter that has a volume that can change and a shape that can change. *page 98*

gene A microscopic part of a living thing that tells the living thing how to develop. *page 17*

germ A microscopic organism, especially bacteria, that can cause disease. *page 30*

geyser A natural fountain of hot water and steam that rises suddenly into the air. *page 79*

glacier A mass of ice that flows slowly over land. *page 66*

graft Take skin from one part of the body and put it over an injured area. *page 31*

gravity The force that causes all objects in the universe to attract each other. It also helps to keep objects in their places. *page 124*

greenhouse effect The warming of the earth caused by an increase of carbon dioxide in the atmosphere. *page 62*

H_2O The chemical formula for water. *page 98*

habitat The place where an animal or plant lives. *page 47*

hairy mammoth A large animal that is no longer alive. Mammoths looked like large, fur-covered elephants. *page 46*

hard water Water that has a high mineral content. *page 110*

heart attack A condition in which part of the heart dies from lack of blood. *page 22*

herbivore An animal that eats only plants. *page 46*

hertz A unit of measurement of pitch. *page 121*

hologram A three-dimensional image made using laser light. *page 134*

hops The dried, ripe flowers of the hop plant used in brewing beer. *page 108*

hot spot A very hot area of the earth's mantle right below the crust. *page 78*

hurricane A violent tropical storm with winds of more than 75 miles per hour. *page 82*

hypothesis A good guess about the answer to a question. *page 17*

iceberg A very large piece of ice floating in the sea. *page 64*

immune system The body's system for fighting disease-causing germs. *page 37*

immunization The process of creating resistance to particular germs. *page 38*

in balance Stable. *page 42*

infection The presence in the body of a disease. *page 30*

insulation A material used to hold in heat. *page 88*

ion An atom or group of atoms that have a positive or negative charge. *page 110*

joint A place where bones are linked together. *page 34*

landfill A place in the ground where garbage is buried. *page 86*

laser A device that gives off light waves that are all exactly alike. *page 133*

lava Melted rock. *page 68*

Law of the Conservation of Matter A rule that says matter cannot be created or destroyed. *page 104*

ligament A tough, elastic band that holds two bones together. *page 33*

light year The distance that light travels in one year. *page 131*

liquid The state of matter that has a volume that can't change but a shape that can change. *page 98*

mantle The part of the earth between the crust and the core. *page 75*

marrow The material inside bones that produces blood cells. *page 33*

mass The amount of matter an object has. *page 124*

matter A substance that occupies space and has mass. *page 94*

mercury A poisonous silver-colored metal that has a low melting point. *page 100*

meteorologist A scientist who studies and predicts the weather. *page 82*

molecule A stable group of two or more atoms. *page 98*

molten Melted. *page 69*

mosses and liverworts Small green plants that grow in damp places. *page 50*

nervous system All the nerves in the body including the brain. *page 25*

neutron An uncharged particle in an atom. *page 95*

newton A unit that measures force. *page 124*

nuclear reactor Where controlled atomic chain reactions take place. *page 139*

nuclear winter A severe drop in temperature resulting from a nuclear explosion. *page 62*

nucleus The center of an atom that contains protons and neutrons. *page 95*

observe Watch and read to gather information about something. *page 16*

optical fiber Thin fibers of glass or plastic used to send information. *page 134*

organ A part of the body that does a particular job. Organs include the heart and the skin. *page 30*

originate Begin. *page 66*

pH scale A measure of how strong an acid or base is. *page 113*

pinwheel A toy made in the shape of a propeller attached to a stick. *page 84*

plankton Microscopic plants and animals that live in the ocean. *page 66*

plate A section of the earth's crust. *page 69*

pollution The dirtying or poisoning of the environment. *pages 53 and 86*

population The number of a group that live in one place. *page 43*

precipitate A solid that forms from a solution due to a chemical reaction. *page 111*

preserve An area set aside for wildlife. *page 43*

prism A piece of clear glass that bends white light, showing the spectrum. *page 130*

proton A positively charged particle in an atom. *page 95*

radar A device that uses radio waves to find and follow storms or other moving objects. *page 84*

recycling Making garbage useful instead of just dumping it. *page 87*

reflection The effect that happens when light, sound, or heat bounces off a wall or other surface. *page 121*

reflex A quick response caused by nerves in the spinal cord. *page 25*

scientific method A process for getting information and testing ideas. *page 16*

seed plants Green plants with roots, stems, leaves, and seeds. *page 50*

shipping lane The official route that ships take between one port and another. *page 66*

soft water Water that has a low mineral content. *page 111*

solar system Our sun and its nine planets. *page 60*

solid The state of matter that has a volume that can't change and a shape that can't change. *page 98*

solution A mixture that consists of one substance dissolved in another. *page 110*

sound wave A pattern of vibrations carried through air, water, and solid objects. *page 120*

species A group of plants or animals that are alike and can produce young together. *page 44*

spectrum The different colors that white light breaks down into: red, orange, yellow, green, blue, indigo, and violet. *page 130*

spinal cord A part of the body that allows messages to travel between the brain and the body. *page 25*

spring A place where water rises out of the ground. *page 79*

state A condition of matter. *page 98*

subspecies A group of plants or animals that look different but can produce young together. *page 47*

symbol One or two letters that stand for an element. *page 96*

tendon A tough, elastic band that attaches a muscle to a bone. *page 33*

theory An idea or set of ideas that tries to explain why certain things happen. *page 79*

tide The rise and fall of the level of the sea. *page 126*

toxin A poisonous chemical. *page 36*

trillion 1,000,000,000,000. *page 131*

tropical Very warm or hot and usually near the equator. *page 83*

tungsten A hard metal that has a very high melting point. *page 100*

universe All existing things, including the earth, the solar system, and the galaxies. *page 105*

vaccine A substance that makes the body produce antibodies against a particular germ. *page 38*

vein A blood vessel that carries blood to the heart. *page 20*

virus A tiny organism, smaller than bacteria, that can reproduce only by using living cells. *page 36*

volcano An opening in the earth's crust through which melted rock is forced. *pages 61 and 68*

water vapor The moisture in the air. *page 60*

wavelength The distance between the top of one wave and the top of the next. *page 130*

white blood cell A cell that can surround and kill germs. *page 37*

yard waste Clippings from lawns and trees, raked leaves, and other plant garbage from yards. *page 87*

yeast A one-celled fungus that breaks down sugar. *page 107*

Answers and Explanations

Check What You Know

Page 7

1. **(B) moves in a circle.** Choice A is not specific enough; the parts have to work in a circular motion. Choices C and D are incorrect because they have nothing to do with circles.

2. **(B) move blood from one place to another.** Choices A, C, and D are not described in the paragraph as having to do with the circulatory system.

3. **(B) without having sexual contact.** Choice A is incorrect because it refers to reproduction with sexual contact. Choice C is incorrect because it is too specific. It describes how the planarian reproduces asexually. Choice D is incorrect because it is not mentioned in the paragraph.

4. **(D) the planarian has some interesting features.** Choices A, B, and C all describe specific features of the planarian. They do not tell the general idea of the whole paragraph.

5. **(A) The tail part would grow a head, and the head part would grow a tail.** Choices B and C are only partly true. Neither part of the planarian would die. Choice D is incorrect because the planarian can grow back parts.

6. **(D) Australia.** Almost all the land in Australia is covered by desert or semi-arid areas.

7. **(C) Europe.** All the continents listed have deserts except Europe.

8. **(D) Africa.** Choice A is incorrect because Africa's desert area is larger than Australia's. Choices B and C are incorrect because neither has a desert as large as Africa's.

9. **(A) salt water.** Choice B is incorrect because it is a single metal, not two metals mixed together. Choices C and D are incorrect because the materials are not dissolved in one another. They are just mixed together.

10. **(C) a penny.** Alloys are solid, and a penny is solid. Choices A and B are incorrect because they involve liquids. Choice D is incorrect because air is not solid. It is a gas.

11. **(D) Both sugar crystals and instant coffee powder would be left in the cup.** The water would evaporate and leave the solids behind. Choice A is incorrect because the solids would be left in the cup. Choices B and C are incorrect because each choice lists only one of the two solids that remain.

12. **(B) 12.** According to the graph, Choices A, C, and D are incorrect because at those ages boys are taller than girls.

13. **(D) between 16 and 18.** The graph shows that boys grow about 5 inches between the ages of 16 and 18. Choices A, B, and C are incorrect because the amount of growth during those periods is less.

14. **(A) Boys and girls would be about the same height.** This is likely because it is the pattern at all ages except age 18. Choice B is incorrect because boys are much taller than girls only when they are full-grown. Choice C is incorrect because girls are never much taller than boys at any point. Choice D is incorrect because it is not true.

15. **(A) flat mirror.** Of the three diagrams, only the diagram of the flat mirror shows light reflecting but not spreading out.

16. **(B) concave mirror.** Of the three diagrams, only the diagram of the concave mirror shows light reflecting toward a center point.

17. **(A) There are three types of mirrors.** Choices B and D are incorrect because they are too specific to be the main idea of the paragraph and diagram. Choice C is incorrect because it is not true. The image seems smaller than the object only in a convex mirror.

18. **(B) burning fuels.** Choice A is incorrect because the paragraph states that burning fuels causes most air pollution. Choice C is incorrect because it refers to only one type of fuel burning. Choice D is a result of air pollution, not a cause.

19. **(C) People would be healthier.** Choices A and B are incorrect because they would be ways to reduce air pollution, not results of reducing air pollution. Choice D is incorrect because the amount of electricity is not necessarily related to the total amount of air pollution.

20. **(A) Air pollution does a lot of damage.** Choices B, C, and D are incorrect because they are all specific aspects of air pollution described in the paragraph.

Unit 1 ◆ Lesson 1

Page 19

1. scientific
2. hypothesis
3. experiments
4. **(A) observation.** Choices B, C, and D are incorrect because they are later steps in the scientific method. Observation of a problem is the first step.
5. **(B) a conclusion.** Choice A is incorrect because an experiment precedes a conclusion. Choice C, a scientific method, refers to the entire process, of which the conclusion is one part. Choice D, an observation, occurs before the experiment is done and the conclusion made.
6. (1) Observe. (2) State the problem as a question. (3) Make a hypothesis. (4) Experiment. (5) Draw a conclusion.

Lesson 2

Page 23

1. vein
2. artery
3. capillary
4. **(A) to the capillaries.** Choice B is incorrect because veins carry blood back to the heart. Choice C is incorrect because arteries carry blood away from the heart. Choice D is incorrect because "used" blood goes to the liver, lungs, and kidneys for cleaning.
5. **(C) to the heart.** Choice A is incorrect because fresh blood passes through the heart and arteries before going to the capillaries. Choice B is incorrect because the veins take "used" blood to the liver, lungs, and kidneys for cleaning. Choice D is incorrect because "used" blood goes to the liver, lungs, and kidneys for cleaning.
6. (1) Eat foods low in fat and cholesterol. (2) Exercise regularly. (3) Stop smoking.

Lesson 3

Page 26
There are many possible ways to answer the question. Share your work with your teacher.

Page 27

1. brain
2. stem
3. cerebrum
4. cerebellum
5. **(C) making sense of what you hear, see, touch, taste, and smell.** Choice A is incorrect because it is an activity of the digestive system. Choices B and D are incorrect because they are activities of the circulatory system.
6. **(D) a quick response.** Choice A is incorrect because a complicated thought does not need to result in quick action. Choice B is incorrect because reflexes take place in healthy people. Choice C is incorrect because a reflex is an action, not a thing.
7. **(B) the bones.** Choice A is incorrect because it is part of the circulatory system. Choice C is incorrect because nerves are part of the nervous system that is being protected. Choice D is incorrect because the eyes receive information for the nervous system but do not protect it.

Pages 28–29
Exercise 1: Circled: Muscles of the Upper Arm. Main idea: Muscles of the Upper Arm.
Exercise 2: Underlined: Body movement, Sense of touch, Speech, Vision, Hearing, Smell
Details: The diagram shows that the cerebrum has several different parts. It also shows that different parts of the cerebrum control different body activities.

Lesson 4

Page 32

1. organs
2. germs
3. cells
4. **(B) keeps the body from drying out.** Choice A is incorrect because bones, not skin, keep the body upright. Choices C and D are incorrect because healthy food and exercise are needed to keep fit and strong.
5. **(B) Burn victims can die from drying out or infection.** Choice A is false, according to the passage. Choices C and D are true, but they are not the reasons bad burns are so serious.
6. (1) Give drugs to fight infection. (2) Replace body fluids. (3) Graft healthy skin over the burned areas.

Lesson 5

Page 35

1. marrow
2. ligament, joint
3. tendon
4. **(A) Bones are made like hollow tubes.** Choices B, C, and D are all things that would make bones heavy, not light.
5. **(B) Minerals and protein fiber are cemented together.** Choice A is incorrect because marrow makes blood cells; it does not add to a bone's

strength. Choices C and D are incorrect because tendons and ligaments serve to attach bones and muscles, not make them strong.

6. There are many possible ways to answer this question. Here is an example.

   There are 2 types of fractures, open and closed. An open fracture is one that breaks through the skin; a closed fracture remains under the surface of the skin.

Lesson 6

Page 39

1. bacteria, viruses
2. toxins
3. Antibodies
4. **(C) to stop the growth and spread of bacteria.** Choice A is what immunization does. Choice B is incorrect because the drugs are not designed to attack the symptoms. Choice D is incorrect because antibiotics do not work against viruses.
5. The cell bursts open.
6. The contents of the cell spill out.

Pages 40–41
Exercise 1: Underlined: But the skin is as important as the other organs.
Exercise 2: There are three kinds of blood vessels: arteries, veins, and capillaries.
Exercise 3: The skin has several jobs.

Lesson 7

Page 44
There are many possible ways to answer this question. Share your work with your teacher.

Page 45

1. ecosystem
2. population
3. preserve
4. **(B) Animals are part of balanced ecosystems.** Choices A, C, and D are incorrect because they are details that support the main idea.
5. **(C) An ecosystem can become unbalanced when a new species is introduced.** Choices A, B, and D are incorrect because they are details that support the main idea.
6. **(A) Their long roots survive fires.** Choices B, C, and D are all incorrect statements not supported by the information in the lesson.

Lesson 8

Page 49

1. Extinct
2. Endangered
3. Carnivores
4. **(C) Tigers were once common, but now they are rare.** Choice A is incorrect because it is not true. Choices B and D are incorrect because they are details that support the main idea.
5. **(A) The jaguar is an endangered species.** Choices B, C, and D are incorrect because they are details that support the main idea.
6. **(D) The earth's climate became colder.** Choices A, B, and C are not supported by details in the passage.
7. **(D) all of the above.** Choices A, B, and C are all part of the many efforts to save endangered animals.

Lesson 9

Page 52

1. algae
2. Ferns
3. Seed
4. **(D) Plants have many uses.** Choices A, B, and C are incorrect because they are examples of the uses of plants. They are details that support the main idea.
5. **(C) leaves.** Choice A, roots, holds the plant in the ground and absorbs water and minerals from the soil. Choice B, the stem, supports the upper part of the plant. Choice D, the flower, produces seeds.
6. **(B) fruit.** Choice A, roots, holds the plant in the ground and absorbs water and minerals from the soil. Choice C, the flower, produces seeds. Choice D, leaves, makes food for the plant.

Lesson 10

Page 55

1. environment
2. pollution
3. DDT
4. **(A) chemical fertilizers.** Choice B is incorrect because the crops themselves don't pollute. Choice C, rainwater, just carries the chemicals into lakes and streams. Choice D has nothing to do with water pollution.
5. **(D) to increase the amount of milk.** Choices A and B are incorrect because just the opposite happens when bGH is used. Choice C is incorrect because the milk is already safe to drink.
6. DDT first entered the food chain when it was sprayed on a mosquito.
7. There are many possible ways to answer this question. Here is an example.

   Once DDT enters a food chain, every member of the food chain is affected.

Pages 56–57
Exercise 1:

1. flowering plant
2. evergreen
3. flowering plant

Exercise 2:

1. carnivore
2. omnivore
3. herbivore

Exercise 3: There are many possible ways to answer the question. Here are some examples of groups.
Plants: rose, tulip
Animals: cat, bluebird, mosquito, dog, fly, eagle
Things with fur: cat, dog
Mammals: dog, cat
Insects: mosquito, fly
Birds: bluebird, eagle
Things that can fly: bluebird, mosquito, house fly, eagle
Things that cannot fly: cat, dog, rose, tulip

Page 58

1. Antibodies, the skin, and vaccination all protect the body against disease germs.
2. There are many possible ways to answer the question. Share your work with your teacher.
3. The loss of the Great Plains, endangered species, and pollution are all caused by human activity and population growth.
4. There are many possible ways to answer the question. Share your work with your teacher.

Unit 2 ◆ Lesson 11

Page 63

1. atmosphere
2. winter
3. greenhouse
4. **(B) helps keep the earth's temperature even.** Choice A is incorrect because the atmosphere protects life on earth. Choice C is incorrect because sunlight does reach the surface of the earth. Choice D is incorrect because the atmosphere extends thousands of miles from earth's surface.
5. **(C) carbon dioxide in the atmosphere.** Although Choices A, B, and D are gases found in the atmosphere, only carbon dioxide increases the greenhouse effect.
6. There are many ways to answer this question. Here is one example. Greenhouse glass allows sunlight to enter the building and then traps the heat inside. This is much like carbon dioxide in the atmosphere because it also does not allow all the heat from the sunlight to escape.

Lesson 12

Page 67

1. current
2. glacier
3. iceberg
4. **(B) Ocean currents have important effects on our lives.** Choices A, C, and D are incorrect because they are details that support the main idea.
5. **(D) The Labrador Current is responsible for the rich Grand Banks fishing grounds.** Choice A is incorrect because it is a detail about the fish in the Grand Banks. Choice B is incorrect because it is a detail about the Grand Banks. Choice C is incorrect because it is a detail that is used to make a transition to the main idea in the new paragraph.
6. **(B) a cold ocean current.** Choice A is incorrect because the only ship mentioned is the Titanic. Choice C is incorrect because the lesson does not mention any glaciers by name. Choice D is incorrect because it describes the Gulf Stream, not the Labrador Current.
7. **(C) northern Europe.** Choice A is incorrect because the Gulf Stream is not in the Pacific Ocean. Choice B is incorrect because that's where the Gulf Stream starts. Choice D is incorrect because the Gulf Stream moves away from South America.

Lesson 13

Page 71

1. eruption
2. crater
3. plate
4. **(C) The volcano rumbled and gave off ash.** Choice A is incorrect because the lesson does not mention wildlife leaving the area. Choice B is incorrect because the weather has nothing to do with predicting an eruption. Choice D is incorrect because lava did not flow before the eruption.
5. **(D) a rough, gray crater.** Choice A is incorrect because a perfect cone shape existed before the eruption. Choice B is incorrect because much of the area is

still bare. Choice C is incorrect because the volcano is not erupting today.

6. **(A) Lava comes out.** Choices B and C are incorrect because the speed of plates and snow are not affected by a volcano erupting.

7. **(D) all of the above.** Mount St. Helens occurs on the North American plate. This plate moves in the opposite direction of, as well as up and over, the Juan de Fuca plate.

Page 73

1. Minneapolis
2. Houston
3. Boston
4. about 500 miles
5. 60s

Lesson 14

Page 77

1. mantle
2. fault
3. earthquake
4. **(B) Predicting where an earthquake will occur is not very difficult.** Choices A and C are incorrect because they are details that support the main idea. Choice D is not mentioned in the paragraph.
5. **(C) Predicting when an earthquake will occur is difficult.** Choices A, B, and D are incorrect because they are details that support the main idea.
6. Most of the faults are located in southern California.
7. There are many answers to this question. Here are some examples: San Francisco, Hollister, Bakersfield, Northridge, and Los Angeles.

Lesson 15

Page 81

1. hot spot
2. geyser
3. **(D) The fact that Yellowstone has geysers suggests that the area is still an active hot spot.** Choices A, B, and C are incorrect because they are details that support the main idea.
4. **(C) a hot spot in the earth's mantle.** Choice A is incorrect because fumaroles produce smoke and gas. Choice B is incorrect because plate collisions occur at the edges of plates. Choice D is incorrect because plates moving apart create volcanoes at plate boundaries.
5. **(B) geysers, hot springs, and fumaroles.** Choice A is incorrect because Yellowstone is on a continent. Choice C is incorrect because Mount St. Helens (the only active volcano in North America) is in Washington State, not Wyoming where Yellowstone is located. Choice D is incorrect because glaciers were not mentioned in the description of Yellowstone.

Lesson 16

Page 84

There are many possible ways to answer this question. Share your work with your teacher.

Page 85

1. hurricane
2. radar

3. tropical
4. **(A) Since 1966 when the first weather satellite was used, hurricanes have not taken us by surprise.** Choices B, C, and D are incorrect because they are details that support the main idea.
5. **(B) 75 miles per hour.** Choices A, C, and D are incorrect because the lesson states in paragraph 3 that winds must be at least 75 miles per hour.
6. **(D) calm area in the middle.** Choices A and B are incorrect because they describe the entire hurricane. Choice C is incorrect because clouds and rain are on the edge of a hurricane.

Lesson 17

Page 88
There are many possible ways to answer this question. Share your work with your teacher.

Page 89
1. Recycling
2. compost
3. landfill
4. **(A) More people are producing more garbage.** Choices B and C are true, but they are not the main reason landfills are being used up. Choice D is also true, but reused metal does not wind up in landfills.
5. **(C) collect materials for recycling at people's homes.** Choice A would encourage people to recycle, but it would not make recycling more convenient. Choice B is incorrect because it would not make recycling easier either.
6. **(B) the dumping of garbage in the ocean.** Choices A and C are ways that we can reduce the amount of garbage in landfills. Choice D is a reason why we must look for other ways to get rid of garbage.
7. **(C) cardboard.** Choices A and B cannot be made of recycled materials. Choice D could itself be recycled, but it not made of recycled materials.

Pages 90–91
Exercise 1: *Spewed* means "thrown out with force." The words around it, such as *exploded* and *violently into the air*, help explain what the word means.
Exercise 2: There are many ways to answer this question. Here is an example. A *radiosonde* is an instrument used to gather weather information for the upper atmosphere.

Page 92
1. Earthquakes and volcanic eruptions such as Mount St. Helens are caused by the movement of plates against one another.
2. There are many possible answers to this question. Here is one example.

   One cycle would be the repeated build-up and release of pressure as earthquakes along the San Andreas fault. Another would be the constant movement of water between the atmosphere and earth.
3. There are many possible answers to this question. Share your work with your teacher.

Unit 3 ◆ Lesson 18

Page 97
1. element

2. compound

3. atomic

4. protons, electrons

5. **(C) two or more elements.** Choice A is incorrect because a compound contains two or more different elements. Choice B is incorrect because the atoms in an element are all alike. Choice D is incorrect because atomic numbers don't make up compounds.

6. **(B) protons and neutrons.** Choices A, C, and D are incorrect because electrons are located outside the nucleus.

7. **(C) a compound.** Choices A and B are incorrect because water is made up of two elements, which could involve more than one element or one atom. Choice D is incorrect because protons alone don't make up a compound.

Lesson 19

Page 101

1. tungsten

2. state

3. liquid

4. **(C) move back and forth.** Choice A is incorrect because the paragraph says that the increase in vibration moves molecules out of place. Choice B is incorrect because nothing in the paragraph mentions breaking down. Choice D is incorrect because the molecules don't change as the ice cube melts.

5. **(D) a tiny wire that lights up.** Choices A and B are incorrect because a filament is not a kind of metal. It is made of metal. Choice C is incorrect because there is nothing in this paragraph about liquid in a light bulb.

6. **(C) lined up in an orderly pattern.** When the H_2O molecules are in an orderly pattern, we call this ice. Choices A, B, and D are incorrect because the paragraph explains that the molecules of ice, water vapor, and liquid water are all the same.

7. **(D) all of the above.** As stated in the paragraph, these states of matter are solid, liquid, and gas.

Pages 102–103
Exercise 1:

1. Mercury freezes at –38° F so the liquid becomes a solid.

2. However

3. Alcohol freezes at a lower temperature than mercury.

Exercise 2:

1. Heat causes the molecules to move faster, so they don't stay in place. This causes the solid to melt.

2. Ice changes from keeping its own shape to taking the shape of its container.

3. The molecules move and become rearranged.

Lesson 20

Page 106

1. universe

2. reaction

3. **(C) the solid matter that is left after something burns.** Choices A and D are incorrect because ash is not a floor or a gas. Choice B is incorrect because the process of burning is a chemical reaction.

4. **(B) to change H_2O to a liquid state (water).** Choice A is incorrect because

it is about a solid. Choice C is incorrect because it talks about a gas and Choice D is incorrect because it has nothing to do with changing states of matter.

5. **(C) Burning changes matter but doesn't destroy it.** Choices A and D are incorrect because they are details related to the main idea. Choice B is incorrect because, when something burns, it changes into other substances. The matter doesn't disappear.

6. **(D) Matter may change state, but it cannot be destroyed.** This is called the Law of the Conservation of Matter. Choices A, B, and C are incorrect because they are part of the main idea.

Lesson 21

Page 109

1. Yeast
2. Fermentation
3. **(A) carbon dioxide gas is produced during fermentation.** The discussion on fermentation states that, as more and more carbon dioxide gas is produced during fermentation, bread dough expands. Choices B, C, and D are all facts given in the lesson, but they are not reasons why bread dough rises.
4. **(D) to make it unfit for drinking.** The last paragraph states that foul-tasting substances are added to keep people from drinking alcohol. Choices A, B, and C are not given in the lesson as reasons why these substances are added to ethanol.
5. **(D) CO_2 and alcohol.** The chemical equation tells us that the breakdown of sugar results in the production of carbon dioxide and alcohol. Choices A and C are incorrect because $C_6H_{12}O_6$, sugar, is not a product of fermentation. Choice B is incorrect because alcohol is only one of these products.
6. **(A) in bread, wine, and beer making.** The lesson does not state that alcohol is produced in the situations described in Choices B, C, or D.

Lesson 22

Page 112

1. solution
2. ions, precipitate
3. soft
4. **(C) calcium stearate.** This substance forms when the ions in dissolved soap and hard water come together. Choices A, B, and D are other substances involved in the process of forming calcium stearate.
5. **(D) Hard water contains more minerals.** Hard water has a high mineral content and soft water has a low mineral content. Choices A, B, and C are not true.
6. **(B) Soap scum forms.** Soap scum forms when soap reacts with calcium ions in the water. Choice A is incorrect because the calcium ions in hard water do not make the water harmful. Choice C is incorrect because the passage states that many people dislike the taste of softened water. The passage does not say that anyone finds the taste of hard water unpleasant.

Lesson 23

Page 115

1. rain
2. base

3. acid

4. **(B) pollutants in the air react with rainwater.** Acid rain forms because pollutants in the air, given off by the burning of fuels, react with gases and rainwater. Choice A is not true. Choices C and D are true, but they are not related to the formation of acid rain.

5. **(C) the killing of plants and animals.** The lesson states that acid rain damages and kills plants and animals. Choices A, B, and D are not included in the lesson as examples of damage caused by acid rain.

6. **(C) North America and Europe.** Choice A is incorrect because South America has no acid rain damage. Choice B is incorrect because Africa has no acid rain damage. Choice D is incorrect because neither South America nor Australia has acid rain damage.

7. **(D) both A and C.** Choice B is incorrect because large amounts of rain aren't necessarily acid rain.

Pages 116–117
Exercise 1:
1. Name, Chemical Formula, Where It Is Found
2. $C_6H_8O_7$
3. hydrochloric acid

Exercise 2:
1. pH, Kind of Animal; pH has a scale.
2. lake trout
3. mussel

Page 118
There are many possible ways to answer questions 1–4. Here are some examples.

1. Chemical reactions cause the wood in my fireplace to burn, the metal on my car to rust, my bread dough to rise, my grape juice to ferment, and the rain that falls to become acid rain.

2. Steam cools to liquid water. Liquid mercury freezes to solid mercury. Liquid oxygen boils to form oxygen gas.

3. The water in my apartment must be hard, because it is difficult to make the shampoo lather when I take a shower. Also, I can see a white layer of scum in the tub.

4. I could help by reducing the amount of fuels I burn. I could do this by using a push lawn mower instead of a gas mower; taking the bus, walking, or riding a bike instead of driving my car; and using a timer on my home's heat thermostat.

Unit 4 ◆ Lesson 24

Page 123

1. decibel
2. Reflection
3. **(D) both A and B.** Choices A and B are incomplete. Choice C is incorrect because the sound waves themselves are invisible.

4. **(A) Sound waves reflect well off walls and other surfaces.** Choices B and C are true. However, they don't explain the reason for an echo. Choice D, therefore, is incorrect.

5. **(B) energy level.** The paragraph is comparing the energy levels of sounds at different decibel levels. Choices A and C are incorrect because pitch and frequency are measured in hertz, not decibels. Choice D is incorrect because *intensity* is a way to describe a sound wave, not another word for *sound* wave.

6. **(D) send.** Choice A is incorrect because the eardrum vibrates, but

it doesn't travel from one place to another. Choice B is incorrect because the eardrum passes the sound waves along; it doesn't make any new waves. Choice C is incorrect because if the waves were absorbed, they couldn't continue on to the inner ear.

Lesson 25

Page 127

1. gravity, centripetal

2. newtons

3. **(C) 80 N.** The graph shows that the force of gravity at 12,800 km above the earth's surface is approximately 80 N. Therefore, Choices A, B, and D are incorrect.

4. **(D) The force is about 675 N more at the surface.** Since that graph shows that the force is 720 N at the surface, about 675 N more than the 45 N at 19,200 km, Choices A, B, and C are incorrect.

5. **(B) The moon would move out into space.** The earth's gravity keeps the moon circling around it. The earth's gravity is an example of centripetal force. Choices A, C, and D are incorrect because they go against Choice B.

Pages 128–129

Exercise 1: How do we hear? <u>The outer ear receives sound waves and directs them toward the eardrum. Like a real drum, the eardrum vibrates and passes the sound waves to the inner ear. The inner ear is full of a liquid that is moved by the sound waves. Hairlike structures in the inner ear feel the moving liquid and send messages to the brain. These messages are what we call hearing.</u>

Exercise 2: (B) They are all about how hearing occurs. Choice C is incorrect because only the facts about the inner ear mention a liquid. Choice A is incorrect because only the last two sentences are about the brain.

Exercise 3: (B) collect sounds so you can hear. Choice A is incorrect because the ear doesn't make sounds. Choice C is incorrect because the ear does not block sound.

Exercise 4:

1. **(B) Objects would fly off into space.** Choices A and C are incorrect because the passage mentions nothing about the speed at which objects would move.

2. A. The paragraph states that we fall down when we trip.

   B. The last sentence states that things such as people, buildings, trees, and the oceans do not fly off into space.

Lesson 26

Page 132

1. prism

2. trillion

3. spectrum

4. wavelength

5. **(B) are alike in some ways.** Choice A is incorrect because both sound and light travel in waves. Choice C is incorrect because two different speeds are given for light and sound. Choice D is incorrect because sound and light travel at different speeds (one difference).

6. **(D) both B and C.** Choice A is incorrect because a prism alone can't cause a rainbow. Choices B and C are incomplete answers.

7. **(D) both A and C.** Choices A and C are incomplete answers. Choice B is incorrect because the light waves appear as separate colors.

Lesson 27

Page 135

1. laser
2. optical
3. hologram
4. **(D) waves that are of the same wavelength and in phase.** Choices A and B are incorrect because they contradict the definition. Choice C is incorrect because white light has many different wavelengths and is not the same as laser light.
5. **(D) both A and C.** Choices A and C are incomplete answers. Choice B is incorrect because reading lamps are not mentioned.
6. **(D) both B and C.** Choices B and C are incomplete answers. Choice A is incorrect because a ruby's atoms do not burn.

Pages 136–137
Exercise 1:

1. You will not see a rainbow. Although there is both sunlight and raindrops, the sun is not behind you so a rainbow won't form.
2. You will not see a rainbow. Clouds on a cloudy day do not allow enough sunlight to pass through to form a rainbow.

Exercise 2: There are many possible ways to answer the question. Here is an example. I would agree with the prediction. Light waves can travel through outer space, but sound waves can't. As a result, the light from the explosion would reach us but the sound would not.

Lesson 28

Page 141

1. chain
2. Fission
3. **(D) both A and B.** Choices A and B are incomplete. Choice C refers to the effects of an atomic bomb, not the addition of a neutron to the atom.
4. **(A) a huge explosion takes place.** Choice B isn't true because the uranium atoms are split, which changes them. Choice C can be a result of the use of this power, but it isn't always a result of the splitting of atoms.
5. **(C) A runaway chain reaction would occur.** Choice A is incorrect because an uncontrolled, not a controlled, runaway chain reaction would most likely occur. Choice B is incorrect because uranium atoms would split very slowly only in a controlled chain reaction. Choice D is incorrect because fission doesn't create new matter.

Lesson 29

Page 144

1. expand
2. contracts
3. **(B) air molecules scatter the shorter wavelengths of light.** Choice A is incorrect because it contributes to the formation of thunder, not colored light. Choice C is the reason

clouds appear white. Choice D is incorrect because Choice C is incorrect.

4. **(C) sound waves from lightning that's farther away from you reaches you later than sound waves from closer lightning.** Choices A and B are incorrect because they are reasons why air vibrates and forms sound waves.

5. **(B) The sky would look brighter red at sunset.** Choice A is incorrect because the dust wouldn't affect the scattering of blue light. Choice C is incorrect because we see more, not less, of the light that is scattered. Choice D is incorrect because there is nothing in the lesson to suggest that the number of clouds would change.

Page 145

1. There are many possible ways to answer the question. Here is an example. I've seen a spectrum "floating" on a pool of oil and created by sunlight shining through a crystal glass and a chandelier.

2. There are many possible ways to answer questions 2–4. Share your answers with your teacher.

Check What You've Learned

Page 146

1. **(B) the water in the river moves more slowly.** Choice A is the opposite of what happens. Choice C is incorrect because the sediment on shore does not cause suspended material to drop. Choice D is incorrect because the suspended material sinks rather than rises.

2. **(A) sand, rocks, and mud that have been deposited at the bottom of a river.** Choice B is incorrect because the material has not been deposited. Choices C and D are incorrect because they do not refer to material deposited by a river or stream.

3. **(C) Every year trees produce a double layer of wood.** Choices A, B, and D all describe the tree's production of layers of wood each year. They are details that support the main idea.

4. **(A) trees grow more during wet years.** The evidence for this is that the rings made during wet years are thicker, meaning that the tree has grown more. None of the other choices is true.

5. **(B) it's 20 years old.** According to the paragraph, a tree produces one pair of rings each year. If the tree has 20 pairs of rings, it must be 20 years old. Choice A is therefore incorrect. Choices C and D may or may not be true, but they do not relate to the age of the tree.

6. **(D) Dallas.** Of the cities listed, only Dallas is covered by the slanted lines that show rain on the map.

7. **(B) in the 20s.** If you look at Chicago on the map, you will see that it is in a clear band area that has temperatures in the 20s.

8. **(A) rainy, in the 50s.** San Francisco is shown covered by the slanted lines that mean rain. It is also in a shaded area that has temperatures in the 50s.

9. **(A) needs more energy to run than a machine that is well oiled.** Choice B is incorrect because machines can have moving parts and friction without having motors. Choice C is incorrect; a machine with a lot of friction costs more to run. Choice D is incorrect because if a machine has a lot of

friction, then it must have parts that rub together.

10. **(B) makes them last longer.** Choices A, C, and D are incorrect because oiling a machine has the opposite effect.

11. **(B) friction wastes energy.** Choice A is incorrect because friction is the result of things rubbing together, not the cause of it. Choices C and D are true, but they are too specific to be the main idea.

12. **(D) atmosphere.** This unit of measure is shown along the bottom axis of the graph. Choice A refers to the space the gas occupies, not pressure. Choice B is not a unit of measure. Choice C is the unit of measure for volume as shown on the left-hand side of the graph.

13. **(B) 1 atmosphere.** You must find the point on the curved line that shows 1 liter. Then you can find the number of atmospheres along the bottom axis that would intersect with this point.

14. **(B) The gas takes up less space.** The graph shows the volume of the gas decreasing (the line slopes down) as the pressure increases. The volume does not increase so Choice A is incorrect. The volume changes, so Choice C is incorrect. Choice D is incorrect because you cannot tell the exact volume of the gas based on the information given in the question.

15. **(B) during the long days of summer.** Choices A and C are incorrect because the diagram shows the geranium does not bloom during short days. Choice D is incorrect because the geranium blooms only when grown during the long days.

16. **(D) September.** The days grow shorter in September, so that's when you would expect to see chrysanthemums. Choices A, B, and C are incorrect because they are all long-day summer months.

17. **(B) long-day plant grown during short days.** Choice A is incorrect because the geranium would bloom if grown during long days. Choices C and D are incorrect because the geranium is a long-day plant, not a short-day plant.

18. **(B) shiny.** The definition of luster is given in the second sentence, after the dash (—). Choices A, C, and D are incorrect because *luster* means "shiny."

19. **(C) electrical wires.** Choices A, B, and D are incorrect because they are all items that should not be good conductors of heat or electricity.

20. **(A) cat.** Of the animals listed, only the cat has a backbone. Therefore, only the cat can be classified as a vertebrate.